THE

PRACTICAL APPLICATION

OF THE

SLIDE VALVE AND LINK MOTION

TO

STATIONARY, PORTABLE, LOCOMOTIVE, AND MARINE ENGINES,

WITH

NEW AND SIMPLE METHODS

FOR

PROPORTIONING THE PARTS.

BY

WILLIAM S. AUCHINCLOSS, C. E.,

MEM. AMER. SOC. CIV. ENG.

TWELFTH EDITION,
REVISED AND ENLARGED.

Slide Valve and Link Motion

by William S. Auchincloss

ISBN 0-917914-13-9

1 2 3 4 5 6 7 8 9 0

PREFACE.

The main object of this Treatise on the Link and Slide Valve, is to place in the hands of the Mechanical Engineer and Draughtsman, a simple method for determining the proportions suitable to any Link Motion, without the assistance of an expensive and cumbrous model, or the delays incident to its manipulation.

Secondarily, to supply the Student of Steam Engineering with a comprehensive view of those causes which regulate both the form and dimension of the cylinder, slide valve and eccentric. This portion of the work has grown incidentally out of the first; for as the link merely combines the action of the two eccentrics, it was obviously necessary that the functions of one of these should be clearly understood before an attempt was made to develop the laws of their joint action. It is hoped, however, that these Parts I. and II. will not prove entirely devoid of interest to the skilled designer, but that they will at least receive a hasty survey, for the sake of the light they throw on the general subject through the medium of Part III.

We are aware, that most Engineers consider the use of a model as absolutely indispensable to the proper adjustment of a Link Motion, and that they are wont to look with skepticism upon all efforts made to solve the problem by other means. So far as these feelings are entertained against an algebraic or trigonometric solution they rest on a firm foundation and receive our hearty approval; for we have been led to believe from careful investigation that it is utterly impossible to construct for this case a formula of any *practical* value, since the question involves from sixteen to twenty variable quantities, the majority of which exert a controlling influence on the motion and many introduce irregularities quite beyond the powers of algebraic expression.

But at the very point where such analysis fails, geometric construction tenders its most efficient aid and furnishes a simple solution of this abstruse problem. The method of investigation has been copiously illustrated with cuts—Part IV—showing the link template drawn in all of its important positions. These have been considered of greater importance for a clear understanding of the subject than the employment of a more diffuse explanation. They are however apt to produce at first sight an erroneous impression regarding the conciseness of the method, but this will always be corrected by a single construction trial, which at once reveals how *small* an amount of work is actually required for accomplishing the object in view.

The Table of crank angles for different locations of the piston cannot fail to effect a great saving of time, for it enables the designer to *instantly* locate the eccentrics for any piston position, without resort to laborious construction.

Besides treating of the various methods by which a link should be suspended for accomplishing certain results, the manner of attaching the eccentric rods has been carefully examined and diagrams presented that show what form should be adopted for meeting most directly the requirements of an accurate adjustment.

A brief discussion of the subject of independent cut-offs will be found in Part V., together with general remarks on Friction, Clearance, Travel, &c.

We have added an Appendix for the benefit of those who desire a mathematical investigation of the subject of crank and piston motion and who would fain inquire more minutely into the principles involved in the STROKE TABLE. This has purposely been separated from the body of the work, which we have aimed as far as possible to preserve free from all algebraic formulæ, in order to render it more acceptable to the majority of mechanics.

We trust that these will appreciate the peculiar *directness* of the solutions effected by the TRAVEL SCALE, in that it determines immediately the very dimensions desired, without taxing the investigator's patience by leading him through intricate constructions to an uncertain result.

The work is now presented to the Engineering community, not without consciousness of its imperfections, yet with the hope that it will tend to simplify in many minds the subject of eccentric and link motion and serve as a stepping-stone to further discoveries.

W. S. A.

NEW YORK. *March,* 1869.

PREFACE TO THE SIXTH EDITION.

THE Author desires to express his gratitude to the Profession for their very cordial reception of his Work, as well as to the several Faculties who have already adopted it, as a text-book in their Colleges.

He has endeavored to eliminate from the present Edition all typographical errors, and otherwise to render it, as far as possible, worthy of regard as a Standard Work in its Department.

NEW YORK, *March,* 1875.

WARNING

Remember that the materials and methods described here are from another era. Workers were less safety conscious then, and some methods may be downright dangerous. Be careful! Use good solid judgement in your work, and think ahead. Lindsay Publications has not tested these methods and materials and does not endorse them. Our job is merely to pass along to you information from another era. Safety is your responsibility.

Write for a catalog or other unusual books available from:

Lindsay Publications
PO Box 12
Bradley, IL 60915-0012

CONTENTS.

PART I.

THE SLIDE VALVE.

ELEMENTARY PRINCIPLES

AND

GENERAL PROPORTIONS.

POWER AND WORK.

THE fundamental query in designing a steam-engine has reference to the power required to accomplish a given amount of work.

The term *work*, when employed in a mathematical sense, signifies the continuous overcoming of an offered resistance along a definite path.

The *quantity of work* is the product of that resistance into the space passed over.

As the standards of weight and distance differ throughout the world, the expressions for quantity of work also differ. With the English standard of pounds avoirdupois and feet, the quantity of work is said to consist of a certain number of *foot-pounds*. But with the French standard of weight, the kilogramme (=2.20462 lbs. avoirdupois) and of distance, the mètre (=3.28089 ft.), the expression becomes a certain number of *kilogrammètres*.

Thus the quantity of work expended in raising a weight of 300 lbs. through a vertical height of 10 ft. =3,000 ft.-lbs. and that of elevating a weight of 50 kilogrammes to a height of 20 mètres =1,000 kilogrammètres. The quantity of work performed by the steam in the cylinder of an engine, equals the mean effective pressure exerted upon the entire area of the piston multiplied by the space passed over in a given time.   The interval of time usually taken is *one min-ute;* hence, if the distance traveled by the piston during a single revolution of the crank be multiplied by the number of revolutions made per minute, their product will equal the required space.

Suppose, for instance, the mean effective pressure on each square inch of a piston, having an area of 1,500 sq. ins., is 60 lbs.; then the total pressure will be $1,500 \times 60 = 90,000$ lbs., and if the crank makes 40 revolutions per minute, with a piston stroke of 3 ft., the speed of the piston becomes 3 ft. $\times 2 \times 40 = 240$ ft. per minute; consequently the quantity of work $= 90,000$ lbs. $\times 240$ ft. $= 21,600,000$ ft.-lbs.

I.—HORSE POWER.

A force capable of raising a weight of 33,000 lbs. one foot high in one minute is termed a Horse power.

The expression originated at the time of the discovery of the steam-engine from the necessity which then arose for comparing its powers with those of the prevailing motor. In its early history this unit had three prefixes—Nominal, Indicated, and Actual—derived from the various methods of estimating the power.   The nominal horse power was based on the general practice of the age, which dealt with low pressures and slow piston speeds.   These quantities have of late years been greatly increased and the old formula in consequence, grown of less and less importance as a true expression of relative capacity.

Indicated horse power designates the total unbalanced power of an engine employed in overcoming the combined resistances of friction and the load.   Hence it equals the quantity of work performed by the steam in one minute, divided by 33,000. Thus, in the above example, the indicated horse power equals

$$\frac{21,600,\cancel{000}}{33,\cancel{000}} = \frac{3)21,600}{11)7,200}\frac{}{654} \text{ HP}$$

The mean effective pressure can alone be determined by means of an instrument called the Indicator.

The Actual or net horse power, expresses the total available power of an engine, hence it equals the indicated horse power less an amount expended in overcoming the friction. The latter has two components, viz: the power required to run the engine, detached from its load, at the normal speed, and that required when it is connected with its load. It is customary in designing *massive engines*—in the absence of reliable data—to estimate the loss of available pressure by the unloaded friction at 2 lbs. per square inch, and subsequently to deduct $7\frac{1}{2}$ per cent. for the friction of the load. Thus, if the mean pressure of the steam within the cylinder.................................... $= 60$ lbs. per sq. in.

It becomes 58 after allowing for unloaded friction, $\dfrac{2}{58}$

And $7\frac{1}{2}\ \%$ of this for the friction of the load....... $= \dfrac{4.4}{53.6}$ lbs.

Gives a net pressure of.....................

But for *small engines* of the ordinary design the total loss by friction will, in many instances, amount to 15 or 20 % of the mean pressure.

Thus, if the mean pressure....................... $= 60$ lbs.

15 % of 60 $=$ total loss by friction .... $= \dfrac{9}{51}$ "

Gives an available pressure of...........

The French apply the term *Force de cheval* to a power capable of raising 4,500 kilogrammes 1 mètre high in 1 minute. Reducing these quantities to their equivalents in pounds and feet and multiplying together, we find that their horse-power equals a force capable of raising 32,549 lbs. 1 foot high in a minute, which is about $\frac{1}{73}$ less than the English unit of measure.

The following TABLE furnishes the Force de cheval equivalents of horse powers ranging between 10 and 100:

Horse Power.	Force de cheval.	Horse Power.	Force de cheval.
10	10.14	60	60.83
15	15.20	65	65.89
20	20.28	70	70.97
25	25.34	75	76.03
30	30.41	80	81.11
35	35.48	85	86.17
40	40.55	90	91.25
45	45.62	95	96.31
50	50.69	100	101.3856
55	55.75	. .	. . .

For powers greater than 100, and less than 1,000, multiply these terms by 10; or, if in excess of 1,000, multiply by 100.

II.—MEAN EFFECTIVE PRESSURE.

The character of the connections between the boiler and steam cylinder, their length, degree of protection, number of bends, shape of valves, etc., must all be considered in forming an estimate of the initial steam pressure in the cylinder; while the mean effective pressure will depend upon the point of cut-off of the steam, and the freedom with which it exhausts.

The exact portion of the stroke that should be completed before this closure or cut-off takes place is a vexed question among engineers, and its discussion is foreign to the object of this Treatise, in which—with the exception of noting certain limits prescribed by different valve motions—it will be considered as predetermined.

Having chosen a point of cut-off, and having estimated the initial pressure of the steam for a given boiler pressure, the question of mean pressure exerted by the steam throughout the piston's stroke, can be approximately solved by the subjoined *Table*, which has been computed in the ordi-

Mean Pressure, Volume, and Temperature Table.

Initial Pressure.	Temperature Fah.	Relative Volume.	STROKE = 1. MEAN PRESSURE FOR VARIOUS CUT-OFFS.						
			$\frac{1}{4}$ or 0.25	$\frac{3}{8}$ or 0.375	$\frac{1}{2}$ or 0.5	$\frac{5}{8}$ or 0.625	$\frac{2}{3}$ or 0.666	$\frac{3}{4}$ or 0.75	$\frac{7}{8}$ or 0.875
Lbs.	Deg.		Lbs.			Lbs.			Lbs.
20	260	765	11.9	14.9	16.9	18.4	18.7	19.3	19.8
25	267	677	14.9	18.6	21.2	23.	23.3	24.1	24.7
30	274	608	17.9	23.3	25.4	27.6	28.	28.9	29.7
35	281	552	20.9	26.	29.6	32.1	32.7	33.7	34.6
40	287	506	23.9	29.7	33.9	36.8	37.3	38.5	39.6
45	293	467	26.8	33.4	38.1	41.3	42.	43.4	44.5
50	298	434	29.8	37.1	42.3	45.9	46.7	48.2	49.5
55	303	406	32.8	40.8	46.6	50.5	51.3	53.	54.4
60	308	381	35.8	44.5	50.8	55.1	56.	57.8	59.4
65	312	359	38.8	48.2	55.	59.7	60.7	62.6	64.3
70	316	340	41.7	52.	59.3	64.3	65.3	67.5	69.3
75	320	323	44.7	55.7	63.5	68.9	69.9	72.3	74.2
80	324	307	47.7	59.4	67.7	73.5	74.6	77.1	79.2
85	328	293	50.7	63.1	71.9	78.1	79.3	81.9	84.1
90	332	281	53.7	66.8	76.2	82.7	84.	86.7	89.1
95	335	269	56.7	70.5	80.4	87.3	88.7	91.6	94.
100	338	259	59.7	74.2	84.6	91.9	93.3	96.4	99.
105	341	249	62.6	77.9	88.9	96.5	97.9	101.1	103.9
110	344	239	65.6	81.6	93.1	101.1	101.6	105.9	108.9
115	347	231	68.6	85.3	97.4	105.6	106.3	110.8	113.8
120	350	223	71.6	89.	101.6	110.2	110.9	115.6	118.8
125	353	216	74.6	92.7	105.8	114.8	115.6	120.5	123.7
130	356	209	77.6	96.4	110.	119.4	120.3	125.3	128.7
135	358	203	80.6	100.1	114.2	124.	125.	130.1	133.6
140	360	197	83.5	103.8	118.5	128.6	130.6	134.9	138.6
145	363	191	86.5	107.5	122.7	133.2	135.3	139.7	143.5
150	365	186	89.5	111.2	126.9	137.8	140.	144.5	148.5
Common difference .			3.0	3.7	4.2	4.6	4.7	4.8	5.0

nary manner with the aid of logarithms (Naperian Base). The first column is given for pressures above that of the atmosphere, or the same as registered by an ordinary steam-gauge. The second and third, for temperature and volume, are taken from Mons. Regnault's Experiments on Saturated Steam. In the estimate for volume, that of the water producing the steam was considered equal to Unity. The Table makes no allowance for clearance.

If from the mean pressure we subtract the mean value of the back pressure, or that which may arise from imperfections in the exhaust, which is usually taken for low-pressure engines at from 1 to 2 lbs. per square inch, the resulting pressure will be the mean effective pressure (in pounds) exerted on each square inch of the piston and may be represented by the letter P.

For high-pressure engines (having an ordinary slide valve) a more exact determination of the mean effective pressure may be secured from the subjoined table, which embodies the results of 50 experiments made by Mr. Gooch, in 1851, with the locomotive "Great Britain," whose boiler pressure varied from 60 to 150 lbs. per square inch.

Mean Effective Pressures incident to a Simple Slide-Valve Motion for various Cut-offs.

Cut-Off at—	Mean Pressure. (Boiler press. $= 1.00$.)	Cut-Off at—	Mean Pressure. (Boiler press. $= 1.00$.)
0.1	0.15	0.45	0.62
$0.125 = \frac{1}{8}$	0.2	$0.5 \ = \frac{1}{2}$	0.67
0.15	0.24	0.55	0.72
0.175	0.28	$0.625 = \frac{5}{8}$	0.79
0.2	0.32	$0.666 = \frac{2}{3}$	0.82
$0.25 \ = \frac{1}{4}$	0.4	0.7	0.85
0.3	0.46	$0.75 \ = \frac{3}{4}$	0.89
$0.333 = \frac{1}{3}$	$0.5 \ = \frac{1}{2}$	0.8	0.93
$0.375 = \frac{3}{8}$	0.55	$0.875 = \frac{7}{8}$	0.98
0.4	0.57	. . . .	. . .

EXAMPLE.

Given $\left\{\begin{array}{l}\text{Boiler pressure} = 70 \text{ lbs. per sq. in.} \\ \text{Steam cut off at } \frac{2}{3} \text{ of the stroke.}\end{array}\right.$

Required.—The mean effective pressure P?

We learn from the table that this pressure for a cut-off of $\frac{2}{3}$ the stroke is 0.82 of the boiler pressure.

$$\text{Then} \quad 70 \times 0.82 = 57.4, \text{ or}$$

The mean effective pressure $P = 57.4$ lbs. per sq. in.

III.—SPEED OF PISTON.

The speed S, or number of feet travelled by the piston in one minute, like the subject of cut-off, rests with the judgment of the individual designer. Nothing more will be attempted in this connection than the presentation of quantities most frequently found in ordinary practice:

Small stationary engines from.......170 to 230 ft. per min.
Large stationary engines...........250 to 300   "
           (Rarely as high as 350 ft.)
River and Sound steamer engines....350 to 500   "
Marine engines.....................250 to 600   "
The Corliss stationary engine........400 to 500   "
           (Usually 50 revolutions.)
Locomotive engines about...........600   "
           (Occasionally 700 or 800 ft.)
The Allen engine...................600 to 800   "
           (Generally the former speed.)

It is interesting to note that a fine specimen of the latter form of engine was operated successfully by Mr. Charles T. Porter, during the late "Exposition Universelle," at the astonishing speed of 1,400 feet per minute.

IV.—DIAMETER OF PISTON.

Having decided the questions relating to indicated horse power, mean available pressure P and piston speed S, all the elements are at hand for determining the area of the piston, and consequently its diameter.

The formula for indicated horse power, solved with reference to such area, will read:

$$A = \frac{33,000 \times \text{Horse power}}{S \times P}$$

or, Area of piston is found by *multiplying the required indicated horse power by 33,000, and dividing the product by speed of piston multiplied by the mean available pressure.*

The corresponding diameter can be obtained from an Area Table.

EXAMPLE.

Suppose that the indicated horse power=100.
Piston speed=300 ft. per minute.
Mean available pressure=21 lbs.
Then the

$$\text{Area} = \frac{33,000 \times 100}{300 \times 21} = 523.8 \text{ sq. in.}$$

Which gives a diameter of about 26 inches.

V.—STROKE OF PISTON.

The general expression for the stroke of an engine (in feet) is,

$$\text{Stroke} = \frac{\text{Piston Speed}}{2 \times \text{No. of Revolutions}}.$$

conversely,

$$\text{No. of Revolutions} = \frac{\text{Piston Speed}}{2 \times \text{Stroke}}.$$

There are many circumstances tending to limit the stroke of a piston. Among other considerations the diameter of a paddle-wheel influences the number of revolutions that can advantageously be made by the crank of a side-wheel steamer, and consequently determines the stroke when the piston speed is chosen. Peculiarities of design frequently make it desirable that an engine should be run at a slow speed and transmit its power through gearing.

Again, the diameters of pulleys for shafting exert an influence, as when the main shaft of a shop is required to run at 120 revolutions per minute, then 60 revolutions for the crank of the engine, will allow a ratio of 2 : 1 between the diameter of the band wheel and shaft pulley.

With a very rapid piston speed, the stroke of the engine is due more to a length imposed on the connecting rod by the necessities of the design, than to the number of revolutions of the crank. In the case of the locomotive, the stroke is generally about 24 inches, and the piston speed 600 feet per minute, while the speed of the engine which depends on its power and the diameter of its drivers, ranges between 20 and 60 miles per hour.

The accompanying table has been calculated, for drivers of different diameters, to represent the number of revolutions they will make per minute, irrespective of slip, when the engine travels at given speeds per hour.

Revolutions made by Driving Wheels of Locomotive at given speeds.

Driving-wheel diameter.	SPEED IN MILES PER HOUR.						Revolutions per mile.
	20 miles.	*25.*	*30.*	*35.*	*40.*	*50.*	
4 ft. 0 in.	140	175	210	. .	. .		420.
4 " 3 "	132	165	198	. .	. .		395.5
4 " 6 "	124	156	186	. .	. .		373.6
4 " 9 "	118	148	177	207	. .		354.
5 " 0 "		140	168	196	. .		336.
5 " 3 "		134	160	187	. .		320.2
5 " 6 "		128	153	179	204		305.9
5 " 9 "		. .	146	170	195		292.3
6 " 0 "		. .	140	163	187		280.3
6 " 3 "		. .	135	157	179	224	269.
6 " 6 "		. .	129	150	172	216	258.6
7 " 0 "		. .	120	140	160	200	240.

(Columns "20 miles" and "50" are each headed vertically "Revolutions per minute.")

The subjoined table is applicable to stationary and marine engines:

No. of Revolutions of Crank for Given Stroke and (approximate) Piston Speed.

Stroke.	PISTON SPEED.														
	Ft. 200	210	220	225	230	240	Ft. 250	260	270	280	290	300	320	340	Ft. 350
1 ft. 6 in.	67	70	73	75	76	80	83	86	90	93	97	100	106	113	116
1 " 8 "	60	63	66	68	70	72	75	78	81	84	87	90	96	100	105
1 " 10 "	55	57	60	61	63	66	68	71	74	76	79	82	88	93	96
2 " 0 "	50	52	55	56	57	60	63	65	67	70	72	75	80	85	87
2 " 3 "	44	47	49	50	51	53	55	58	60	62	64	66	72	76	78
2 " 6 "	40	42	44	45	46	48	50	52	54	56	58	60	64	68	70
2 " 9 "	36	38	40	41	42	43	45	47	49	51	53	55	58	62	64
3 " 0 "	33	35	36	37	38	40	42	43	45	47	48	50	53	56	58
3 " 3 "	31	32	33	34	35	37	38	40	41	43	44	46	50	52	54
3 " 6 "	29	30	31	32	33	34	36	37	38	40	41	43	46	48	50
3 " 9 "	27	28	29	30	31	32	33	34	36	37	39	40	43	45	47
4 " 0 "	25	26	27	28	29	30	31	32	34	35	36	38	40	42	44
4 " 3 "	23	24	25	26	27	28	29	30	32	33	34	35	38	40	41
4 " 6 "	22	23	24	25	26	27	28	29	30	31	32	33	35	38	39
4 " 9 "	21	22	23	23	24	25	26	27	28	29	30	31	33	36	37
5 " 0 "	20	21	22	22	23	24	25	26	27	28	29	30	32	34	35

These dimensions, the stroke of piston and diameter of cylinder, are so constantly used in comparing engines of different powers, that, as far as possible, they should consist of whole numbers quite free from all fractions of an inch.

VI.—AREA OF STEAM PORT.

This dimension ranks next to cut-off in its controlling influence upon the proportions of the valve seat and face. It may justly be considered as a *Base* from which all the other dimensions are derived in conformity with certain laws. Its value depends greatly upon the manner in which the port is employed, whether simply for admitting the steam to the cylinder, or for purposes both of admission and exit. In cases of admission it is evident that the pressure will be sustained at substantially a constant quantity by the flow of steam from the boiler. But in cases of exit or exhaust, a limited quantity of steam, impelled by a constantly *diminishing* pressure, forces its way into the atmosphere with less and less velocity. If, then, the engine is supplied with two steam and two exhaust passages, the ports will be correctly proportioned when the areas of the latter *exceed* those of the former by an amount indicated by careful experiment. When, however, one passage performs *both* duties, it should have an area suitable for the exhaust and be opened only a limited amount for the admission of the steam. Very excellent results have been found to attend the employment of an area equal to 0.04 of that of the piston, and a steam-pipe area of 0.025 of the same, when the speed of the piston does not exceed 200 ft.

per minute, but widely-different factors are demanded by higher speeds, like those peculiar to locomotives.

In the year 1844 M. M. Gouin and Le Chatelier instituted a series of experiments for ascertaining the value of such terms. These were continued about six years later by Messrs. Clark, Gooch, and Bertera, upon engines of British manufacture. The various results having been collated and analyzed by Mr. Clark, were finally presented to the public in his valuable work on "Railway Locomotives." From this it appears that with a piston speed of 600 ft. per minute, an area of 0.1 that of the piston was found to give practically a perfect exhaust, a steam-pipe area of 0.08 a free admission of steam to the chest, and a port opening of from 0.6 to 0.9 the entire width of the port, depending on the humidity of the steam, a free admission to the cylinder.

The following table has been prepared for intermediate speeds of the piston on the assumption that for average lengths of pipe the area increases as the speed, and that a higher speed is usually attended by increased pressure:

Speed of Piston.	Port Area.	Steam-Pipe Area.
200 feet per minute.	.04 area of piston.	.025 area of piston.
250 " "	.047 " "	.032 " "
300 " "	.055 " "	.039 " "
350 " "	.062 " "	.046 " "
400 " "	.07 " "	.053 " "
450 " "	.077 " "	.06 " "
500 " "	.085 " "	.067 " "
550 " "	.092 " "	.074 " "
600 " "	.1 " "	.08 " "

Having determined the area of the steam port, the next step will be to resolve it into its factors, length and breadth. When a small travel of the valve is essential, the length should be made as nearly equal to the diameter of the cylinder as possible; then the port area divided by the length, furnishes of course the value of the breadth or S in Fig. 1.

The extent to which the valve should open this port for the admission of the steam will equal from 0.6 to 0.9 of the value of S, and the *minimum travel* of the valve, that which with a given cut-off just opens the steam port the amount of this limit. The *maximum travel* is governed by expediency, the general tendency of an excess over the minimum travel is to render the events of the stroke *more* decisive, the cut-off takes place with greater brevity, avoiding unnecessary wire drawing of the steam and the release opens rapidly, affording a more perfect exit. Where the travel is small, these good qualities should be secured by increasing the travel, until the valve gives an opening equal to or even greater than the width of the steam port. With a large travel no such attempt should be made, since it would inevitably sacrifice much work in friction and cause a far greater loss than gain.

EXAMPLE.

Diameter of a certain piston=26 inches. Area=531.

Piston speed=350 ft. per minute.

Required.—Width of steam port, minimum width of port opening and diameter of supply steam pipe.

From the Tables we have :

Sq. inches.

Area of steam port=$531 \times .062 = 33$ sq. inches.

The length of the port=diameter of cylinder=$26''$.

And the width=$\frac{33}{26}=1.3$ inches or $1\frac{5}{16}$.

Minimum width of port opening=$0.6 \times 1.3 = \frac{3}{4}$ inch.

Sq. inches.

Area of steam pipe=$531 \times .046 = 24.4$ sq. inches.

Consequently the diameter=$5\frac{1}{2}$ inches.

In the Corliss Engine, where the steam is admitted and exhausted through different valves, it is customary to give the steam passage an area of $\frac{1}{15}$ to $\frac{1}{16}$ that of the piston, and the exhaust an area of from $\frac{1}{10}$ to $\frac{1}{11}$.

In this connection a few remarks may appropriately be made with reference to the formation of the valve edge and the walls of the steam port. The experiments of scientists like Weisbach, D'Aubuisson and Koch, prove that the various phenomena of contraction in the fluid vein observed in the flow of water are equally true for gases, the formulæ of discharge however have slightly different coefficients of efflux. The character of the discharge will evidently vary with the extent of opening offered by the valve edge, from what is termed "discharge through a thin plate" at the commencement, to that through a "short tube" with the full opening. Fig. 1 illustrates the natural convergence

FIG. 1.

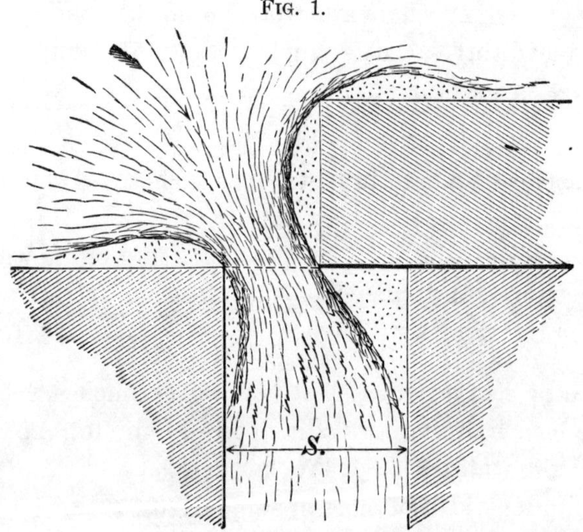

which takes place in the filaments of the steam vein with the common slide valve. If the edge were formed as in Fig. 2 the discharge would be much improved and rendered similar to that which occurs through an ordinary "mouth piece."

The curvature of the valve edge should commence **far**

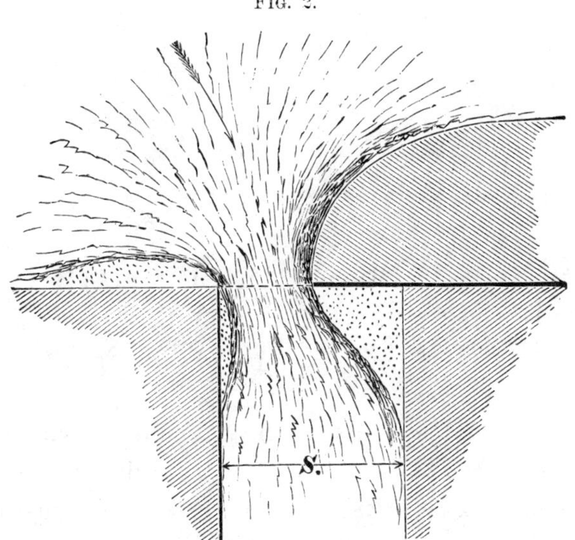

enough above the rubbing surfaces to permit a limited amount of wear without altering the proportion of the parts.

Every effort should also be made to reduce the amount of clearance for the steam and loss of head by friction, to a minimum value. Hence the passage from the port to the cylinder must be constructed as short as possible, be of uniform cross section and bend with easy curves if bending is indispensable.

In the moulding of a cylinder casting, the cores for the steam and exhaust passages should be faced with very great care, in order to secure surfaces along which the steam will flow with perfect freedom.

PISTON, CRANK

AND

VALVE MOTIONS.

In essaying the study of an intricate subject like the relative motions of the piston and the ordinary slide valve of a steam engine, it is of the utmost importance to first divest the parts of all the complicating influences which arise from special constructions and present them in such simple and elementary forms, that the discovery of the fundamental laws governing their motions may be facilitated. If these are clearly defined, the deduction of others adapted to special cases will subsequently be accomplished with comparative ease.

The entire series of events which take place within the cylinder of an engine, occur when the piston has reached definite positions in its complete stroke. It follows (since there is in practice no fixed limit to the stroke) that an infinite number of such positions may be occupied, and in order to express them by a standard which shall apply equally to all cases, a unit scale must be adopted. The stroke of all pistons therefore will be regarded throughout this Treatise as equal to *Unity*, and their positions at certain important periods, as decimal portions of the entire stroke.

If a movable point is caused to travel around a fixed

one, in the same plane, at a constant distance therefrom, it will describe a curved line called a circle. For the purpose of locating any position in the path of the movable point, the circle has from remote ages—though not wisely—been divided into 360 equal parts called degrees (360°), each degree into 60 minutes and each minute into 60 seconds.

While the piston of an engine performs a single stroke, the crank-pin makes a semi-revolution (180°) about the centre of the main shaft, each position of the former consequently corresponds with some angular position of the crank-arm, and if these angles are arranged in a Table we can instantly determine therefrom the number of degrees over which the pin must pass in order to bring the piston to any desired position.

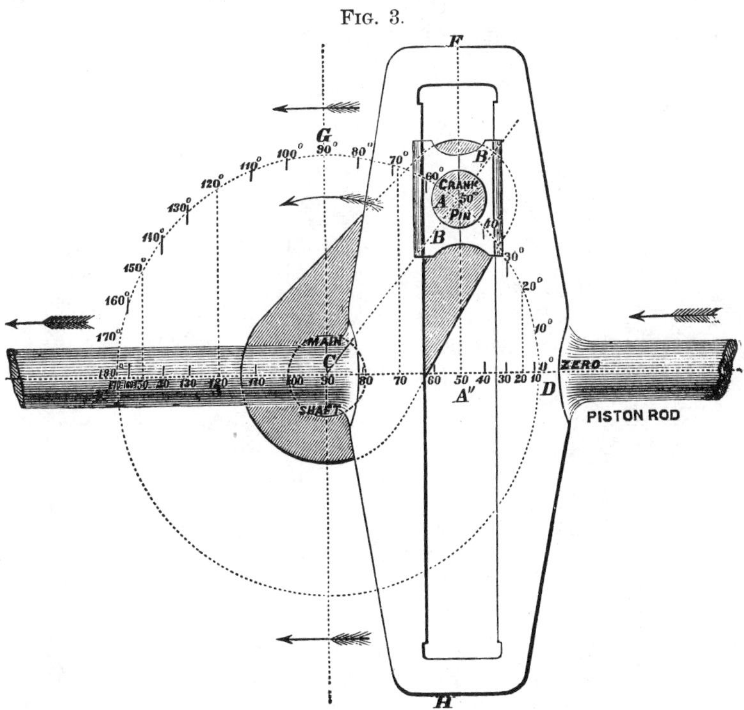

Since the "slotted cross-head" shown in Fig. 3 is the only form of connection between the crank-pin and piston, in which the piston moves from one extremity of the stroke to the other at the same speed as the crank-pin—measured on the stroke line—it will answer our purpose for determining the fundamental principles of the piston and valve motions. The arrangement of the parts are clearly shown in the Figure. The crank-pin is surrounded by blocks BB, these slide freely up and down the solid frame FH to which the piston-rod is welded, so that while the crank-pin advances from D to G the block mounts towards F, returns as it approaches E and descends towards H on the return stroke ED. For convenience, the cylinder will always be regarded as lying on the *right-hand* side of the main shaft and the point of the crank-pin circle *nearest* to the cylinder as the zero or starting point of the forward stroke.

TABLE A.

Piston Position.	Crank Angle.	Piston Position.	Crank Angle.	Piston Position.	Crank Angle.
	Deg.		Deg.		Deg.
0.1	$36\frac{7}{8}$	$0.5625 = \frac{9}{16}$	$97\frac{1}{8}$	$0.813 = \frac{13}{16}$	$128\frac{5}{8}$
$0.125 = \frac{1}{8}$	$41\frac{3}{8}$	0.575	$98\frac{5}{8}$	0.82	$129\frac{3}{4}$
0.15	$45\frac{5}{8}$	0.6	$101\frac{1}{2}$	0.83	$131\frac{1}{4}$
0.175	$49\frac{1}{2}$	$0.625 = \frac{5}{8}$	$104\frac{1}{2}$	0.84	$132\frac{7}{8}$
0.2	$53\frac{1}{8}$	0.65	$107\frac{1}{2}$	0.85	$134\frac{3}{8}$
0.225	$56\frac{5}{8}$	$0.666 = \frac{2}{3}$	$109\frac{1}{2}$	0.86	$136\frac{1}{8}$
$0.25 = \frac{1}{4}$	60	0.68	$111\frac{1}{8}$	0.87	$137\frac{3}{4}$
0.275	$63\frac{1}{4}$	$0.687 = \frac{11}{16}$	112	$0.875 = \frac{7}{8}$	$138\frac{5}{8}$
0.3	$66\frac{3}{8}$	0.69	$112\frac{3}{8}$	0.88	$139\frac{1}{2}$
0.325	$69\frac{1}{2}$	0.7	$113\frac{5}{8}$	0.89	$141\frac{1}{4}$
$0.333 = \frac{1}{3}$	$70\frac{1}{2}$	0.71	$114\frac{7}{8}$	0.9	$143\frac{1}{8}$
0.35	$72\frac{1}{2}$	0.72	$116\frac{1}{8}$	0.91	$145\frac{1}{8}$
$0.375 = \frac{3}{8}$	$75\frac{1}{2}$	0.73	$117\frac{3}{8}$	0.92	$147\frac{1}{8}$
0.4	$78\frac{1}{2}$	0.74	$118\frac{5}{8}$	0.93	$149\frac{3}{8}$
0.425	$81\frac{3}{8}$	$0.75 = \frac{3}{4}$	120	0.94	$151\frac{5}{8}$
$0.437 = \frac{7}{16}$	$82\frac{7}{8}$	0.76	$121\frac{3}{8}$	0.95	$154\frac{1}{8}$
0.45	$84\frac{1}{4}$	0.77	$122\frac{5}{8}$	0.96	$156\frac{7}{8}$
0.475	$87\frac{1}{8}$	0.78	$124\frac{1}{8}$	0.97	$160\frac{1}{8}$
$0.5 = \frac{1}{2}$	90	0.79	$125\frac{1}{2}$	0.98	$163\frac{3}{4}$
0.525	$92\frac{7}{8}$	0.8	$126\frac{7}{8}$	0.99	$168\frac{1}{2}$
0.55	$95\frac{3}{4}$	0.81	$128\frac{1}{4}$	1.00	180

The foregoing Table furnishes angular positions of the crank-arm corresponding with the various points in the stroke which may at times be occupied by the piston.

To illustrate its application, suppose for—

EXAMPLE.

The stroke of a certain piston=36 inches.

Query.—How many degrees will the crank have passed over when the piston reaches points respectively $9''$ and $23\frac{3}{8}''$ distant from the commencement of its stroke?

1st.
$$\frac{9''}{36''}=6)\overline{1.50}=0.25 \text{ of the stroke.}$$
$$6)\overline{9.00}$$
$$\overline{0.25}$$

2d.
$$\frac{23\frac{3}{8}}{36}=\frac{23.38}{36}=0.649 \text{ of the stroke.}$$

Then by the Table:

    0.25 of the stroke=an angular passage of 60°.

    0.65   "   =   "   " $107\frac{1}{2}$°

The required angles.

AGAIN: Suppose the stroke of a piston=$36''$, and that the crank has passed over 112°. How far will the piston have advanced?

The Table gives for 112° a piston position of 0.687 of the stroke.

Therefore $0.687 \times 36''=24\frac{3}{4}''$ the distance advanced by the piston while the crank has advanced 112 degrees.

There is securely fastened to the crank shaft a device called an "eccentric," which serves to impart a reciprocating motion to the slide valve. Upon close inspection it appears that this is only a mechanical subterfuge for a *small crank.*

The travel of any valve being small compared with that

of its piston, the crank required for its motion has frequently an arm or "*throw*" *c b* shorter than one-half the diameter *a e* of the main shaft, Fig. 4.   Hence to avoid cut-

FIG. 4.

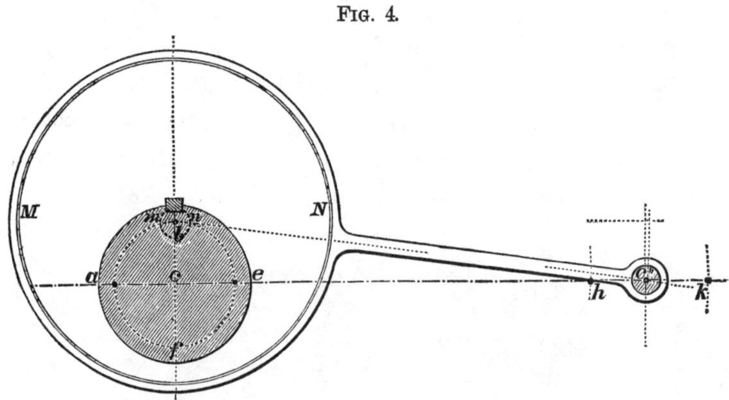

ting the shaft and the expense of forming the crank *c b*, the pin *m*, *n*, and enclosing strap of the rod are greatly enlarged until they attain the common diameter M N, the former may then be slipped on, and keyed fast to the shaft *a e*.   Of course the motion will not be altered by this change, but the same reason that led to the adoption of the slotted cross head for tracing the piston's progress, now compels us to substitute a small slotted cross head and rod for the eccentric rod.   In the sequel therefore both the crank pin and the eccentric pin (or centre of eccentric) will be considered as transmitting their motions through slotted cross-heads to the piston and the valve.   (See Fig. 5.)

The axes of the cylinder and of the valve stem do not always pass through the centre of the main shaft.   When that of the latter lies above and parallel to the former, as shown in the figure, some expedient must be adopted for carrying the motion of the eccentric pin up from the point *q* in the central plane of the engine to *e* in that of the valve.

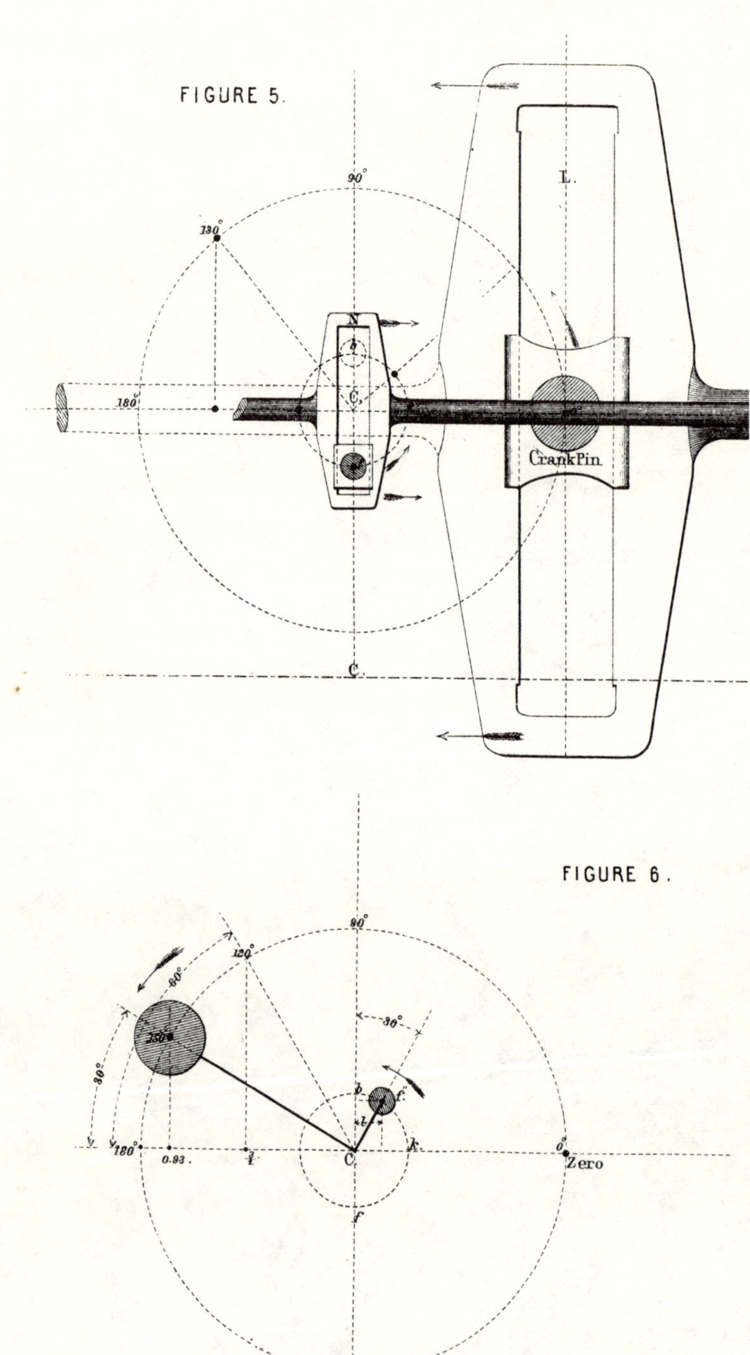

FIGURE 5.

FIGURE 6.

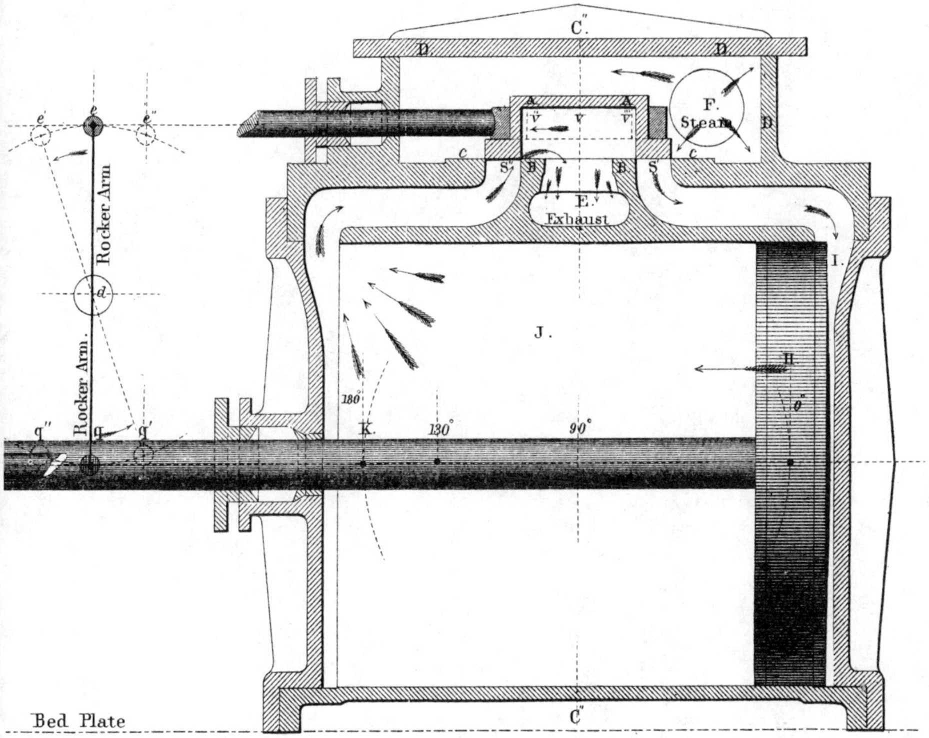

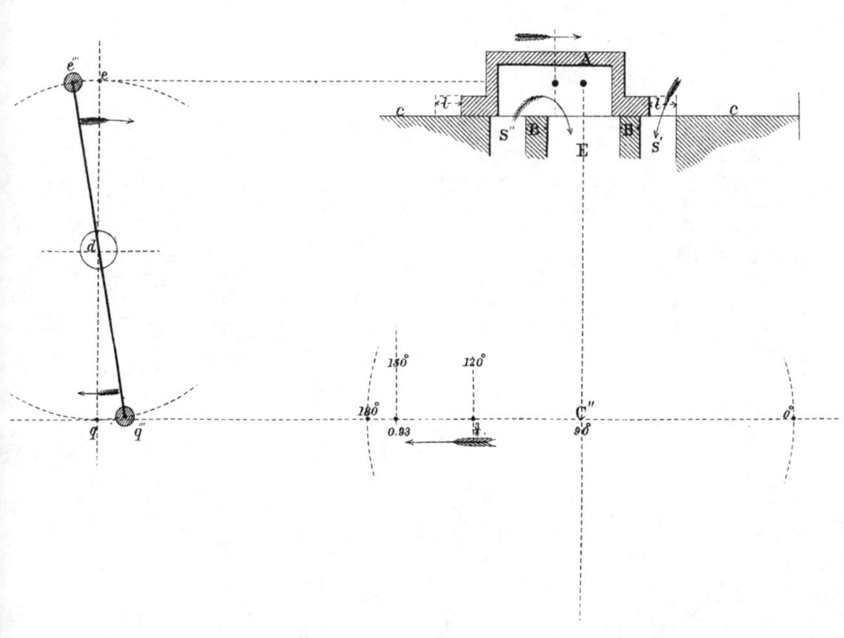

Rocker Arm
Rocker Arm.
e
e'
e''
d
q''
q
q'
Bed Plate
C''
D
D
A
A
F.
Steam
D
v
v
v'
c
c
S
B
E
Exhaust
B
S
I.
J.
H
θ''
180
K.
130°
90°
C''
e'''
e
d
q
q''
A
c
v
v'
c
S''
B
E
B
S'
S
150°
120°
180°
C''
0°
0.93
90°

This is frequently accomplished by aid of a bar $q\,d\,e$ called a "*rocker*," free to oscillate on its firmly supported axis d. The direction of the motion then becomes the reverse of that produced by the eccentric pin and if the pins q and e are made to operate in vertical slots no irregularity will be introduced by this arrangement.

Having explained the general features of these controllers of motion, the crank and the eccentric, and having resolved them into their elementary forms, we pass to consider the parts moved and seek the law of their proportions.

The plain slide valve of a steam-engine is a device by which the entrance and exit of the steam is regulated for the opposite ends of the cylinder. It is essentially a case A, resting on a plane surface $c\,c$ as seen in cross section in Figs. 5 and 11. Through this surface are cut three passages S', S'', and E, separated by the partition walls B, B, called "*bridges*." The two former lead to the opposite extremities of the cylinder, and the passage E called the "*exhaust*" leads through an oval pipe to the atmosphere. The valve A is sufficiently large to cover both the passages S', S'', when standing in its neutral position. A second case D, D, called the "*steam chest*," encloses the valve A and is secured rigidly to the plane surface $c\,c$. Being larger than the valve it leaves over it much unoccupied space to which the only entrance is through the aperture F. This space is the "reception room"—so to speak—of the cylinder ; to it, the steam is admitted from the boiler through F and kept in waiting during such times, as the valve in its motion completely covers the two ports.

Figure 5 represents the crank-pin at the zero point of its path, the piston at the extremity H of its stroke, the valve in the neutral position and all the parts ready for motion. A complete revolution of the crank will carry the piston forward to K and return it to the starting point H. Whatever events take place in the journey from H to K should be repeated in the same order on the return route from K to H, hence in studying the motion we will seek to render it perfect for the trip from H to K and leave the parts when the latter point is reached in the same relative positions as those occupied for H, so that the one will become simply a counterpart of the other. The first point evident, is that the port S′ must be opened and again closed for the proper admission of the steam during the stroke of the piston from H to K; in other words, while the piston is making one entire stroke the valve must accomplish a half and a return half of its stroke. Such an operation can only be brought about by securing the eccentric pin in the position f or b on a line at right angles to the crank-arm, that of f being suitable for a direction of the crank indicated by the arrow.

Let us trace the two motions throughout one revolution of the crank. Moving it from the zero to the 90° point will draw the piston from the position H to the half stroke or the line c'', c'', will advance the eccentric pin from f to k, the rocker from $e\,q$ to $e'\,q'$, the centre of the valve from V to V″ and completely open the port S′. As the crank progresses from 90° to 180° the eccentric pin will travel from k towards b, gradually closing the port S′ and completely covering it when the 180° point is reached, thus leaving the valve in the same position at the terminus of the stroke that it occupied at the commencement. On the return stroke from K to H the port S″ will in like manner be opened and again closed. In thus hastily following the two entrances of the steam to the cylinder, we have lost sight of its mode of escape after performing the work of forcing along the piston. Let us suppose that one revolution has been completed and the piston is prepared for a second journey from the position H. The space J is now filled with steam and some passage of escape must be opened. This is provided in the port and pipe E, which are thrown into immediate communication with the passage S'' when the valve commences its motion, the opening becoming wider and wider as the travel progresses, only closing when the piston reaches the point K and is ready to receive fresh steam through the passage S'' for the return stroke.

Such is a brief outline of the parts and functions of the simplest form of slide valve, in which the steam is admitted at the commencement of the piston's stroke and not excluded until that stroke is completed.

This arrangement, however, is not attended with economic results, for it entirely ignores that remarkable property of steam, its elasticity. To render this latent power available, the steam should be admitted during only a portion of the piston stroke, the valve should then be closed and the confined volume of steam allowed to complete the remaining portion, by developing its power of expansion.

But how can our elementary form of valve and position of eccentric be modified for attaining this desirable result?

Suppose a cut-off were required at a piston position of 0.93 of the stroke. By carrying the crank to the 150° position (as in Fig. 6) we observe that the port S remains opened a distance l and the most ready means for effecting its closure is to lengthen the valve face by this amount. Since the cut-off must take place at relatively the same piston position in both strokes, an equal addition must be made to the other edge of the valve. Such additions to the outer edges of the valve, for the purposes of cut-off, are called overlap or simply "*lap*." The extent of this lap in the present case is evidently equal to the horizontal distance of the eccentric's centre f'' from the 90° line, because without lap it would naturally close at this line.   The same distance expressed in degrees would be equivalent to a "*lap angle*" of 30°.

But on referring to Fig. 6 it is clear that no such addition can be made without necessitating a change also in the eccentric location, for it would render the admission 30° *too late*.   Hence if we add a lap to the valve equivalent to an eccentric motion of 30° from its neutral position, we must at the same time unkey the eccentric, and having advanced it also 30° refasten it on the main shaft.   The number of degrees by which the eccentric is thus carried forward from a position at right angles to the crank-arm is termed the "*angular advance*" of the eccentric.

When the eccentric stands at right angles to the crank the exhaust closes and release commences at the *extremities* of the stroke, consequently if the eccentric be moved ahead 30° not only will the cut-off take place 30° earlier, or at a crank angle of 120° instead of 150°, but the release as well as the exhaust will take place 30° earlier or at the 150° crank angle.   Although we have not secured by this process the cut-off aimed at, yet the investigation distinctly points out the means at our command for the accomplishment of *any* cut-off and will enable us to construct a Scale for determining the magnitudes of such alterations.   For a cut-off of 140° there would be required an angular advance of 20° and a lap equivalent to the distance these degrees remove the eccentric centre from the line at right angles to the crank ; for a cut-off of 160°, an advance of 10° with a corresponding lap, and so on ; the exhaust closure taking place respectively at the 160° and 170° crank angles.

This closure of the exhaust confines the steam in the cylinder until the port is again opened for the return stroke ; consequently the piston in its progress will meet with increasing resistance from the steam which it thus compresses into a less and less volume. Such opposition when properly proportioned aids in overcoming the momentum stored up in the reciprocating parts and tends to bring them economically to a state of rest at the end of each stroke. Since the closure of one port is simultaneous with the opening of the other, a release will take place of the steam which was previously impelling the piston. Within certain limits this also is conducive to a perfect action of the parts, for an early release enables a greater portion of the steam to escape before the return stroke commences, whereas a release at the end of the stroke would be attended by a resistance of the piston's progress, from the simple fact that steam *cannot* escape *instantaneously* through a small passage, but requires a certain definite portion of time dependent on the area of the opening and the pressure. The larger the opening then the less the occasion for anticipating the moment of exhaust.

We learn therefore that the moments of exhaust closure and release are, when the valve has neither "*inside lap*" nor its converse "*inside clearance*," directly dependent upon the angular advance of the eccentric, and that an angular advance of 20° produces a closure at a crank angle of 160°, one of 30° at 150° and so on, the resistance becoming continually greater as the angular advances increase. A limit at length is reached where this resistance really becomes detrimental, and an amount of power is absorbed quite inconsistent with economy of action. On this account the single eccentric is rarely used to effect cut-offs of less than $\frac{2}{3}$ the stroke. Earlier cut-offs require two valves and two eccentrics, the one set for regulating the cut-off of the steam, the other its admission and escape. This subject will be more fully discussed in Part V.

The principles just developed can be embodied in a single Diagram called the TRAVEL SCALE, whose construction is illustrated by Fig. 7.

FIG. 7.

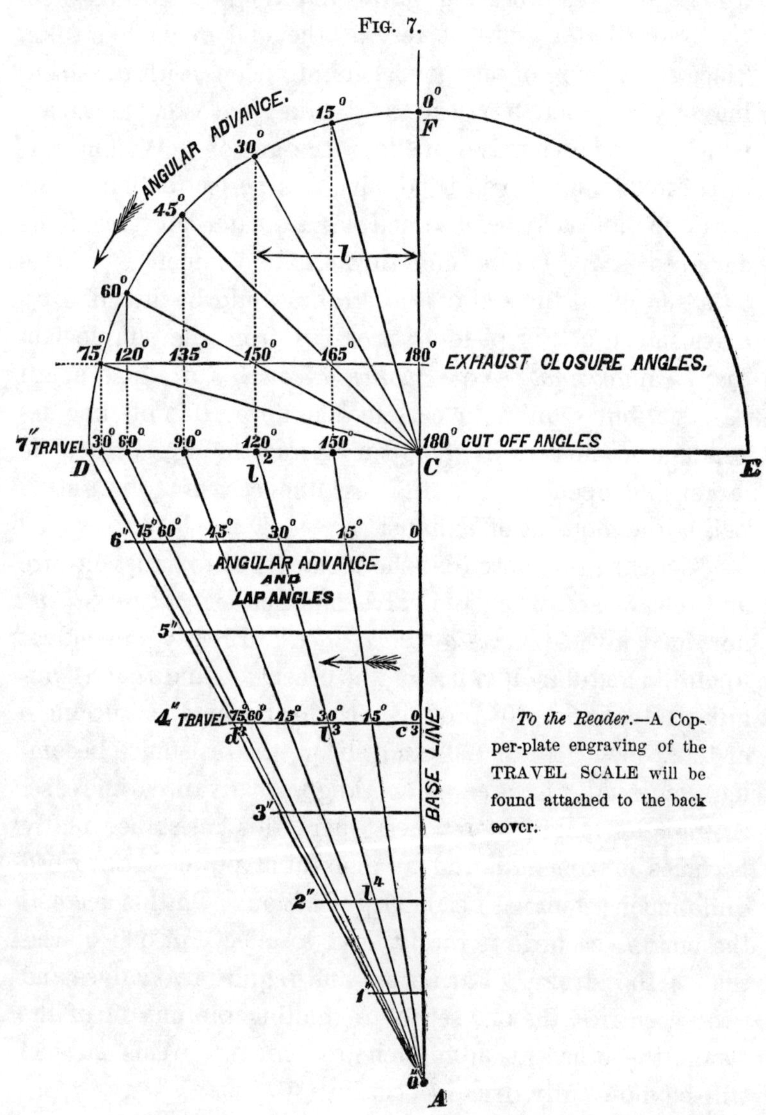

Let E F D represent the path traversed by the centre of an eccentric whose throw equals $3\frac{1}{2}$ inches, consequently the travel of its valve$=7$ inches.   Then C F at right angles to D E will be the normal position of the eccentric from which the angular advances must be laid off.   Extend this line to some convenient point A and join the extremity D of the travel with A.   Divide the line C A into 7 equal parts, and through these points draw lines parallel to D E to represent all the travels less than 7 inches.   Finally project each degree of the arc D F upon the line D C and join the points thus found with the point A.

The distances from the Base Line C A, at which this group of lines intersect the travel lines, will indicate what lap should be given to accomplish various cut-offs, and their distances from the extreme travel line D A will give the width of the steam-port opening due to these travels and cut-offs.   Thus for $7''$ travel and a cut-off of $120°$ the eccentric must have an angular advance of $30°$ and the valve a lap equal to l^2 C, giving thereby a port opening l^2 D; while a travel of 4 inches with the same cut-off only requires a lap of l^3 C^3 and has a port opening of l^3 d^3.   The exhaust closure of course takes place in both cases at a crank angle of 150, or piston position of 0.93 the stroke.

It will be observed that this Scale may be applied with perfect accuracy to travels greater than 7 inches by making these lines represent their multiples; for instance, a 4-inch travel may stand for one of 8 or 12 inches; a 6-inch travel for one of 12 or 18 inches, and so on.   In such cases the values of the true lap and lead will be double or thrice those given by the Scale.   Since the same principle holds for travels less than 2 inches, it is clear that the Scale must apply to all possible dimensions.

A slip of paper and a pencil are the only paraphernalia of the Travel Scale.   To illustrate its use take for—

EXAMPLE.

Extreme width of port opening must$=1\frac{1}{4}$ inches and the valve must cut off steam at 0.82 the stroke.

Required.—Angular advance of the eccentric, travel of valve, lap and point of exhaust closure.

Table A gives for a piston position of 0.82 the stroke a crank angle of 130°, for this cut-off an angular advance of 25° will be required (see line C D of the TRAVEL SCALE).

Apply the edge of a slip of paper to the Inch Scale and mark off the desired width of the port opening a, b, as in

FIG. 8.

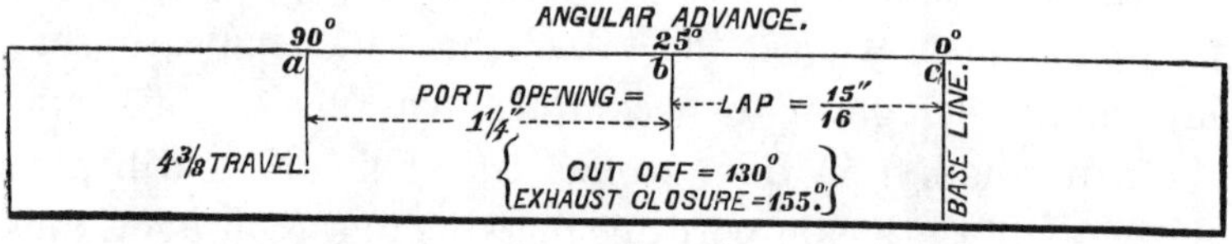

ANGULAR ADVANCE.

Carry the same to the Travel Scale, place the mark a over the 90° line C A and slide the edge—parallel to the line C D—until the mark b stands directly over the 25° angular advance or lap-angle line. The $4\frac{3}{8}$ inches line of travel, upon which the slip of paper here stops, will be the correct travel for the valve. Before removing the paper mark the position c of the Base line. Finally return the slip to the Inch Scale and measure the lap $b\,c$, which gives $\frac{15}{16}$ of an inch. The exhaust closure on one side and release on the other will of course take place at the 155° angle of the crank (see line C D) or at a piston position of about 0.95 of the stroke.

ANSWERS.
{ Angular Advance$=25°$.
Travel of valve$=4\frac{3}{8}$ inches.
Lap$=\frac{15}{16}$ inch.
Exhaust closes at 0.95 of the stroke.

The solution of such problems as the subjoined, will

tend to familiarize the Reader with the method of using
this Travel Scale :

1st. To cut off at $\frac{2}{3}$ the stroke, with port opening of $1\frac{1}{8}$
inches.

Required.—Angular advance of the eccentric, travel of
valve, lap and point of exhaust closure.

2d. To cut off at $\frac{3}{4}$ the stroke, with port opening of $1\frac{5}{8}$ ins.

3d.   "   " $\frac{7}{8}$ "   "   "   "   "   3 "

4th.   "   " 0.7 "   "   "   "   " $\frac{13}{16}$ "

DIRECTION OF CRANK MOTION.

The direction of any crank motion depends on two conditions—1st. The presence or absence of a rocker for transmitting the motion; 2d. The location of the angular advance with reference to the central line of the valve motion.
Both of these may be conveniently expressed in a single
Diagram like the accompanying Fig. 9, in which the posi-

Fig. 9.

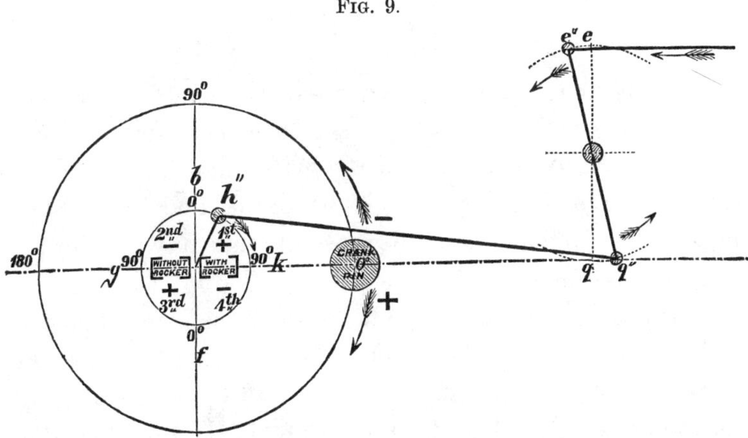

tive sign (+) represents a motion in the direction of the
hands of a watch, the negative (−) a reverse motion.   To

produce a positive motion in any engine, whose eccentric acts through a rocker, lay off the angular advance from the line $b\,f$ in the 1st quadrant (the crank standing at the zero), but for one without a rocker, the angular advance must be laid off from the same line in the 3d quadrant. The 4th and 2d quadrants in like manner belong to the negative motion. The reason for making such a disposition of the angular advance will at once appear upon tracing out either of these motions.

When the power of an engine is transmitted through a wide belt to the machinery, the direction of its crank motion will be determined by the relative locations of the main and crank shafts. The strain should *invariably* be made to fall upon the *lower* portion of the belt, the upper being thereby relaxed, sags upon its pulleys, increases the frictional surface, and materially improves the adhesion of the belt.

LEAD.

This term is applied to an alteration made in the plan of the valve motion for the purpose of concealing and neutralizing an effect, due to imperfect workmanship as well as continual wear in the boxes of the crank and cross head pins. The difficulty may be best explained with the assistance of Fig. 10.

Suppose, for instance, both boxes of the connecting rod A B, fit loosely upon the crank and cross-head pins, that the crank moving in the direction indicated by the arrow, has reached a location C A within 8 degrees of the zero, and that the piston (on account of the lost motion in the boxes) falls short of its true position B, a distance B B. If now the

momentum of the motor carries the crank-pin past its zero,
the piston, which at the moment of passage is no longer
urged or restrained by the connecting rod, will by virtue

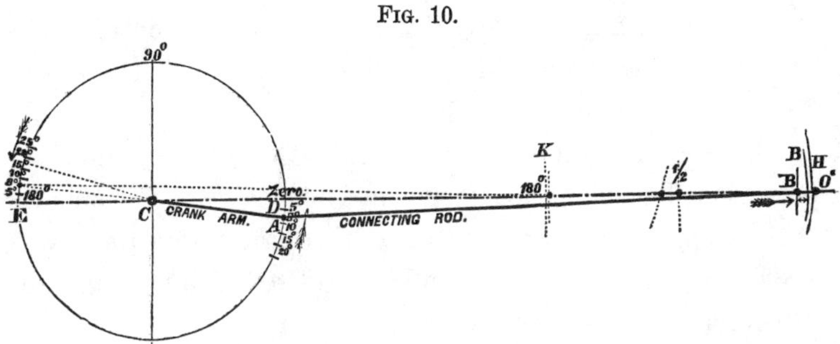

FIG. 10.

of its own momentum continue moving in the direction of
H until all the lost motion being expended, its progress is
suddenly checked and it is itself again brought under the
control of the connecting rod, which then draws it forward
upon the return stroke. These concussions are reproduced
at the end of each stroke with a degree of force and sound
directly dependent on the extent of the lost motion and the
momentum of the piston with its connecting rod. Where
the parts are of great weight, as in a marine engine, the
sound becomes very loud and the engine is said to
"*thump*" or "*pound*" on the "*centres*." Two ways pre
sent themselves for counteracting this effect; the one, by
making the boxes so durable and the workmanship so perfect that lost motion becomes almost impossible; the other,
by introducing a resistance to the momentum of the piston
capable of *completely* overcoming it before the end of the
stroke, in other words by allowing the steam to enter the
cylinder a short time *previous* to the termination of the
stroke. With small engines the first method is practicable,
but in large ones both are more commonly employed be-

cause with these, a very small amount of lost motion suffices to produce a disagreeable sound.

The width of port opening given by any valve at the moment its crank passes either centre, is called the "*lead*" of the valve ; and the angular distance of the crank from its zero at the instant this opening commences, the "*lead angle.*"

The opening together with the angle (or time) limit the power of the steam in its effect upon the lost motion ; for even a small opening continued through a long time may prove as efficient for the admission as a large opening during a very short time.

Since sound, the effect of lost motion, depends upon the weight and velocity of the reciprocating parts, the lead requisite must vary for different engines and also for the same engines at different velocities. The *exact* amount *cannot* be predicated in any particular case, but after the engine has been constructed it may be experimentally determined by gradually increasing the angular advance of the eccentric until some position is found which results in a smooth and noiseless movement of the reciprocating parts. We have before alluded to the effect of compression by a premature closure of the exhaust, but it must be distinctly understood that this agency unassisted cannot neutralize the evils of lost motion without injuring the admission of the return stroke. In this respect it differs from lead. It should then usually be supplemented by lead in order to accomplish a smooth action of the parts and free opening of the steam port for the return stroke. Observe also, that so long as the lead angle amounts to only a few degrees no impression can be produced on the continuity of the crank motion, for the lever arm will be too small for the power to exert any influence over the crank.

The limits of the *lead angle* are commonly *zero and* 8° *for stationary engines;* while for any given angle the width of opening will depend upon the travel of the valve and the point of its cut-off.

It remains to be shown that the Travel Scale is quite as applicable to valves having a certain lead as to those without any. Referring again to Fig. 6, imagine an increase in its angular advance of 5°, the valve will then close at 115° instead of 120° and reopen its port 5° *before* the crank reaches the extremity of the stroke; but if the lap be reduced 5° when the angular advance is increased 5°, the cut off will still remain 120°, while the port commences to open 10° *before* the end of the stroke. Consequently if we wish to arrange a valve for a certain number of degrees lead, without altering the point of cut-off, it will simply be necessary to *find the angular advance for a valve without lead, add ½ the lead angle for a new angular advance, and subtract the other ½ for an angle by which to measure the lap.*

If in the Example of Fig. 8 a *lead of* 8 *degrees* had been required, with the *same* cut-off, the angular advance

$$\left\{ \begin{array}{l} \text{would have become } 25° + 4° = 29° \\ \text{and the lap angle } \quad 25° - 4° = 21° \end{array} \right\}$$

and by applying the port opening marks a and b to the 90° and 21° lines,—instead of the 90° and 25° lines,—we would have obtained a travel of $3\frac{7}{8}$ inches and a lap of $\frac{11}{16}$ inch; while the distance between the angular advance line of 29° and the lap angle line of 21° would have equaled $\frac{1}{4}$ inch, the *width* of the *lead opening* at the extremity of the piston stroke.

The change in the angular advance of course changes the exhaust closure from 155° to 151° or about 0.93 of the piston's stroke.

Supposing then a lead angle of 8° for the same problem the answers become :—

Angular Advance........................... $=29°.$
Lap angle................................ $=21°.$
Travel................................. $=3\tfrac{7}{8}$ inches.
Lap.................................. $=\tfrac{11}{16}$ inch.
Lead................................. $=\tfrac{1}{4}$ inch.

Exhaust closes at 0.93 of the stroke.

Similar suppositions made and applied to the other trial problems will give all the practice requisite for successfully using the TRAVEL SCALE.

It seems almost unnecessary to observe that the SCALE effects with equal readiness and precision solutions directly the converse of that just accomplished. Thus, if the above lap, lead and travel were given, to determine the exhaust closure and cut-off, we would mark the lap and lead on a strip of paper as in Fig. 12, apply the same to the $3\tfrac{7}{8}$ inch travel line of the scale, which would show at once an angular advance of 29° and consequently exhaust closure of 0.93 the stroke ; also a lap angle of 21° with lead, or 25° without, the same as a cut-off of 130°=0.82 the entire stroke.

A moment's reflection will also show that—during the progress of the crank—the varying width of the port opening from the simple lead out to the maximum width and back again to the period of cut-off, might *readily* be traced on the SCALE, and all the information common to the popular method of ellipse or other construction, be immediately obtained. But the facts thus gained, would prove of very *trifling* moment, so long as the valve had received a correct maximum port opening.

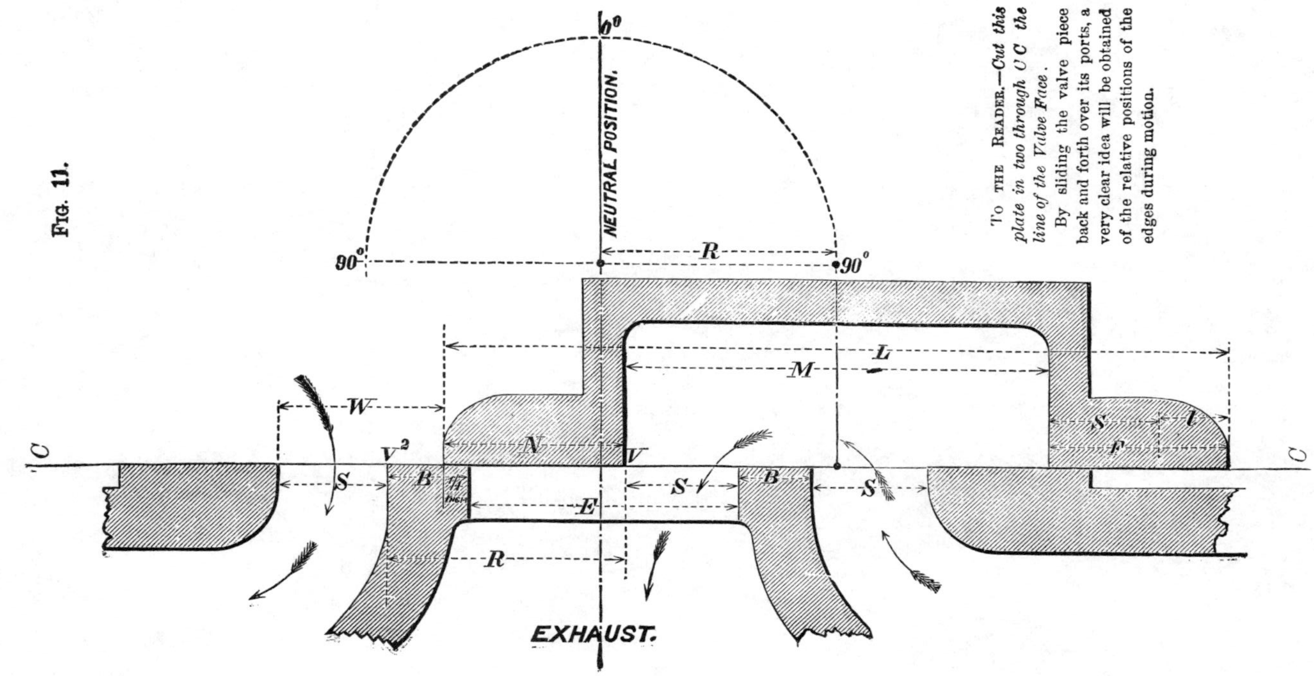

FIG. 11.
NEUTRAL POSITION.
EXHAUST.
To the Reader.—Cut this plate in two through C C the line of the Valve Face.

By sliding the valve piece back and forth over its ports, a very clear idea will be obtained of the relative positions of the edges during motion.

WIDTH OF BRIDGE.

This dimension is usually made of equal thickness with the cylinder, in order to secure a perfect casting, but at times it becomes necessary to increase its width. The only danger from a narrow bridge is an *overtravel* of the valve, by which the exhaust passage would be placed in direct communication with the "live steam" in the chest, and followed by continual waste of the power. Obviously this cannot occur while the difference between the port opening and the steam port does not exceed the width of the bridge. (Fig. 11.) But to prevent even the possibility of a leakage :—

Add about ¼ of an inch to the width of the opening and from their sum subtract the width of the steam port.

Thus the width of the steam port in the example of Fig. 8, should have been at least :—

$$1\tfrac{1}{4} + \tfrac{1}{4}'' - 1'' = \tfrac{1}{2} \text{ inch.}$$

When however the width of the opening is less than that of the steam port, the danger of such an escape entirely vanishes.

WIDTH OF EXHAUST PORT.

The main difficulty to be avoided in proportioning the width of this port is the possibility of a reduction in its area, when the valve attains extreme travel, to an opening materially less than that of the steam port from which it derives its supply.

Suppose that the valve in Figure 11 has reached the end of its half travel, or the exhaust edge V moved a distance

R from its neutral position V^2; then by the above condition, E will evidently equal (S+R−B).

Which furnishes the following general

RULE

For determining width of Exhaust port.

Add the width of the steam port to ½ the travel and from their sum subtract the width of the bridge.

When called upon to perform the addition or subtraction of many fractional portions of an inch, it will generally be found more convenient to express these decimally than by those *very awkward* subdivisions sixty-fourths, thirty-seconds, etc.

Fractions of an inch expressed decimally.

Fraction	Decimal	Fraction	Decimal	Fraction	Decimal
$\frac{1}{64}$ of inch	=.0156	$\frac{1}{4}+\frac{3}{32}$ of inch	=.3438	$\frac{5}{8}+\frac{1}{16}$ of inch	= .6875
$\frac{1}{32}$ "	=.0313	$\frac{3}{8}$ "	=.375	$\frac{5}{8}+\frac{3}{32}$ "	= .7188
$\frac{1}{16}$ "	=.0625	$\frac{3}{8}+\frac{1}{32}$ "	=.4063	$\frac{3}{4}$ "	= .75
$\frac{3}{32}$ "	=.0938	$\frac{3}{8}+\frac{1}{16}$ "	=.4375	$\frac{3}{4}+\frac{1}{32}$ "	= .7813
$\frac{1}{8}$ "	=.125	$\frac{3}{8}+\frac{3}{32}$ "	=.4688	$\frac{3}{4}+\frac{1}{16}$ "	= .8125
$\frac{1}{8}+\frac{1}{32}$ "	=.1563	$\frac{1}{2}$ "	=.5	$\frac{3}{4}+\frac{3}{32}$ "	= .8438
$\frac{1}{8}+\frac{1}{16}$ "	=.1875	$\frac{1}{2}+\frac{1}{32}$ "	=.5313	$\frac{7}{8}$ "	= .875
$\frac{1}{8}+\frac{3}{32}$ "	=.2188	$\frac{1}{2}+\frac{1}{16}$ "	−.5625	$\frac{7}{8}+\frac{1}{32}$ "	= .9063
$\frac{1}{4}$ "	=.25	$\frac{1}{2}+\frac{3}{32}$ "	=.5938	$\frac{7}{8}+\frac{1}{16}$ "	= .9375
$\frac{1}{4}+\frac{1}{32}$ "	=.2813	$\frac{5}{8}$ "	=.625	$\frac{7}{8}+\frac{3}{32}$ "	= .9688
$\frac{1}{4}+\frac{1}{16}$ "	=.3125	$\frac{5}{8}+\frac{1}{32}$ "	=.6563	1 *inch*	=1 000

INSIDE LAP.

The effect on a valve motion of inside lap is to—
Prolong the Expansion, and
Hasten the Compression.
(A contrary effect for inside clearance.)

The former is occasionally added in the case of high-speed engines having very late cut-offs. In such instances the compression is arranged to commence at about $\frac{7}{8}$ of the stroke, or at an angle of 138 degrees, and the release at an angle not exceeding 160°. For example, if the angular advance equals 32° (with a travel of $4\frac{5}{8}$ inches) the compression would commence at a crank angle of 148° or 10° later than the above limit; hence if we give the valve an inside lap of 10° or $\frac{3}{8}$ of an inch found as in Fig. 12, the expan-

FIG. 12.

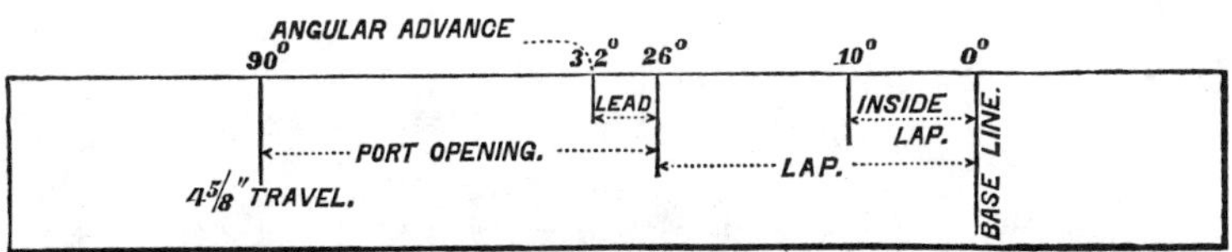

sion will continue from the point of cut-off to $148° + 10° =$ 158 degrees, and the compression commence at $148° - 10° =$ 138 degrees, instead of both events taking place at the 148° angle of the crank.

We think the foregoing investigations fully sustain our remarking in conclusion that any questions, relating to the travel of the valve, the varying widths of the exhaust and steam-port openings for every possible position of the crank, the moments of closure and release, and other points of interest, can not only be determined with *perfect* precision by means of the TRAVEL SCALE, but their solution will prove well nigh *instantaneous* when compared with the indirect and tedious methods that have heretofore obtained in popular usage.

4

GENERAL EXAMPLE.

What dimensions should be given to the cylinder and valve of an engine like Fig. 5 to secure an indicated horse power of 150 with

Pressure of steam in boiler at 65 lbs. ;

The crank to make 50 revolutions per minute, and the steam to be cut off at $\frac{2}{3}$ the stroke?

The mean effective pressure (page 16)$=65 \times 0.82 = 53.3$ lbs.   Piston speed (page 17)$=$say 250 ft. per minute.   Area of piston, page 18,

$$A = \frac{33,000 \times 150}{250 \times 53.3} = \frac{132 \times 150}{53.3} = 371 \text{ sq. inches.}$$

Therefore diameter of piston$=21\frac{3}{4}$ ins., say 22 inches.

Stroke of piston (page 19)$=\dfrac{250}{100}=2.5$ ft.$=2$ ft. 6 inches.

Port area (page 22)$=371$ sq. inches $\times .047 = 17.4$ sq. inches.

If the length of the steam port$=20$ inches then its width will$=\dfrac{17.4}{20}=\frac{7}{8}$ inch.

Width of port opening W (by page 23) may vary between 0.6 and 0.9 the width of the entire port, but for the sake of greater precision in the cut-off and freer opening of the port at the commencement of the stroke, let us make its width equal about 1.5 width of steam port, or—

$$W = 1.5 \times \tfrac{7}{8}'' = 1\tfrac{1}{4} \text{ inches.}$$

Area of steam pipe (page 22)$=371$ sq. inches $\times .032 = 11.9$ square inches.

Area of exhaust pipe$=$area of steam port$=17.4$ sq. ins.

The respective diameters of these pipes will therefore be 4 and $4\frac{3}{4}$ inches.   By the Travel Scale, the angular advance for the given cut-off of 110° equals (without lead) 35° and with a lead angle of say 6°,

Angular advance will$=35°+3°=38$ degrees,
And lap angle will$=35°-3°=32$ degrees.

Now apply the width of port opening $1\frac{1}{4}$ inches to the 90° and 32° lines of the Travel Scale, as page 38, and we find that the Travel must$=5\frac{3}{8}$ inches.

After marking the Base line and angular advance we have—

Lap$=1\frac{7}{16}$ inches ; lead$=\frac{1}{4}$ inch.

The bridge, page 47, should not be less than $\frac{1}{4}''+\cdot1\frac{1}{4}''-\frac{7}{8}''$ $=\frac{5}{8}$ inch. If, however, the cylinder has a thickness of 1 inch the bridge must be made of the same width.

Width of exhaust port, page 48,

$$E=\tfrac{7}{8}+2\tfrac{3}{4}-1''=2\tfrac{5}{8} \text{ inches.}$$

Also we have the width of each valve face F and N=width of steam port+lap

Equals, $\tfrac{7}{8}''+1\tfrac{7}{16}''=2\tfrac{5}{16}$ inches,

And the total length of the valve or

L=exhaust port + 2 bridges + 2 faces$=2\frac{5}{8}+2+4\frac{5}{8}=9\frac{1}{4}$ inches.

The angular advance being 38° the exhaust will close and release commence at the 142° angle of the crank (see Travel Scale) or at 0.895 of the stroke$=30''\times0.895=26\frac{7}{8}$ inches and the cut-off take place at $\frac{2}{3}\times30''=20$ inches; which embraces all the required dimensions.

PART II.

GENERAL PROPORTIONS

MODIFIED BY

CRANK AND PISTON CONNECTION

CRANK AND PISTON CONNECTION.

Thus far we have confined our attention to a form of connection called the "slotted cross-head," and have been able therewith to deduce laws governing the proportions of the various parts of the valve, as well as to devise a most simple and rapid method for determining their magnitudes. But since this connection seldom obtains in practice, it becomes necessary for us to analyze the form shown in Fig. 13, to modify their general proportions to accord with the new conditions and to eliminate as far as possible all the irregularities they tend to create.

It will be observed, by inspecting this Figure, that the cross head pin is drawn a distance BB″ beyond its half stroke position B″, when the crank attains an angle of 90°, that this irregularity is due to the want of parallelism of the connecting rod, with its original position—during the progress of the crank pin in its semi-revolution—and that a rod of virtually infinite length produces a motion of the piston identical with that of the cross-head. It follows that the irregularity BB″ will vary with the different ratios that may exist between the length of the crank arm and the connecting rod. In subsequent comparisons of these two terms, the length of the *crank arm* will always be regarded as the *Unit measure* and that of the connecting rod as a certain number of times the length of the crank arm.

Let the crank arm C A be equal to unity and the connecting rod A B=4, then their ratio is that of 1 to 4, (1 : 4.) When the arm occupies the 90° position the cross-head pin will be drawn a distance B B″ beyond the half stroke point

Fig. 13.

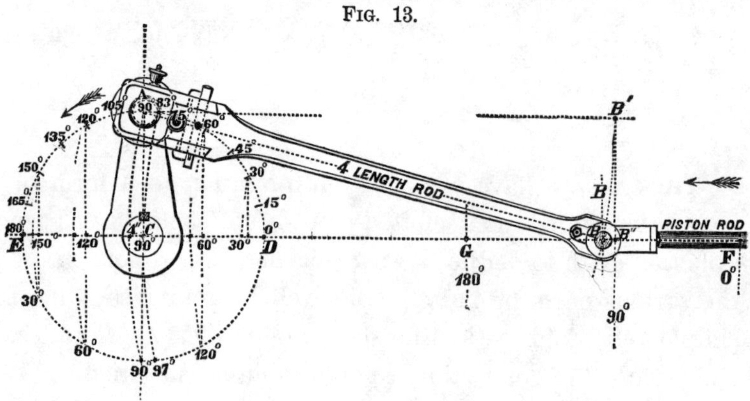

B″. With B as a centre and A B as radius, describe the arc A A″. If the occasion required, it might be readily proved that A″, the point of its intersection with the line D E, is the same distance from C that B is from B″. Placing the crank in other positions—as at 30°, 60°, 120° and so on —and describing similar arcs there will result like irregularities but of a less degree, all of which however vanish at the extremities of the stroke D and E. It becomes evident therefore that the effect of this form of connection is ; *to carry the piston ahead of its proper positions throughout the forward stroke and on the return stroke to make it lag behind the positions due to the locations of the crank pin.*

Consequently the *one crank angle*, for a given piston position (as in Table A), will no longer serve both the forward and return strokes, but a *new* table must be constructed which shall furnish at sight the proper angles of the crank for various piston positions in both the Forward and the Return strokes, and these for every important ratio of crank to connecting rod between 1 : 4 and 1 : 8 with which intermediate values may readily be determined by interpolation.   Such is presented in the following STROKE TABLE.*

The fractional portions of a degree have been given as small as can conveniently be laid off with a protractor.

By transposing the terms *Forward* and *Return* the angles in the Table will apply to the case of a "*Back Action*" *Engine.*   For the irregularities of the motion are necessarily *reversed* in such instances, because the cross head and cylinder lie on *opposite* sides of the main shaft instead of on the *same* side.

FIRST EXAMPLE.

The connecting rod of a certain engine=8' 3"=99".
The crank arm=18 inches.
Cut-off takes place at 0.65 of the stroke.

Required—The forward and return stroke crank angles.

Divide length of connecting rod by that of the crank arm : thus

$$\frac{99}{18}=5\tfrac{1}{2}$$

Their ratio therefore will be that of 1 : $5\tfrac{1}{2}$.†

---

* The trigonometric formulæ pertaining to this Table are furnished in the Appendix and angles, for any special case may be found by their aid.

† A Ratio commonly adopted by the builders of stationary engines.

STROKE TABLE.

Piston Position. (Stroke = unity.)	CRANK ANGLES. (FOR ORDINARY CONNECTING ROD.)								
	RATIO 1 : 4.			RATIO 1 : 4½.			RATIO 1 : 5.		
	Forward	Return.	Diff.	Forward	Return.	Diff.	Forward	Return.	Diff.
	deg.	deg.	deg.	deg.	deg.	deg.	deg.	deg.	deg.
$0.125 = \frac{1}{8}$	$37\frac{3}{8}$	$46\frac{3}{4}$	$9\frac{3}{8}$	$37\frac{5}{8}$	$46\frac{1}{4}$	$8\frac{5}{8}$	$37\frac{7}{8}$	$45\frac{5}{8}$	$7\frac{3}{4}$
0.2	48	$59\frac{1}{2}$	$11\frac{1}{2}$	$48\frac{1}{2}$	$58\frac{3}{4}$	$10\frac{1}{4}$	$48\frac{3}{4}$	$58\frac{1}{8}$	$9\frac{3}{8}$
$0.25 = \frac{1}{4}$	$54\frac{3}{8}$	$66\frac{3}{4}$	$12\frac{3}{8}$	$54\frac{7}{8}$	66	$11\frac{1}{8}$	$55\frac{3}{8}$	$65\frac{3}{8}$	10
0.3	$60\frac{3}{8}$	$73\frac{3}{8}$	13	61	$72\frac{5}{8}$	$11\frac{5}{8}$	$61\frac{1}{2}$	72	$10\frac{1}{2}$
$0.333 = \frac{1}{3}$	$64\frac{1}{4}$	$77\frac{5}{8}$	$13\frac{3}{8}$	$64\frac{3}{4}$	$76\frac{7}{8}$	$12\frac{1}{8}$	$65\frac{3}{8}$	$76\frac{1}{4}$	$10\frac{7}{8}$
$0.375 = \frac{3}{8}$	$68\frac{7}{8}$	$82\frac{3}{4}$	$13\frac{7}{8}$	$69\frac{1}{2}$	82	$12\frac{1}{2}$	$70\frac{1}{4}$	$81\frac{3}{8}$	$11\frac{1}{8}$
0.4	$71\frac{3}{4}$	$85\frac{5}{8}$	$13\frac{7}{8}$	$72\frac{3}{8}$	$84\frac{7}{8}$	$12\frac{1}{2}$	73	$84\frac{1}{4}$	$11\frac{1}{4}$
0.45	$77\frac{3}{8}$	$91\frac{1}{2}$	$14\frac{1}{8}$	$78\frac{1}{8}$	$90\frac{5}{8}$	$12\frac{1}{2}$	$78\frac{5}{8}$	90	$11\frac{3}{8}$
$0.5 = \frac{1}{2}$	$82\frac{7}{8}$	$97\frac{1}{8}$	$14\frac{1}{4}$	$83\frac{3}{4}$	$96\frac{1}{4}$	$12\frac{1}{2}$	$84\frac{3}{8}$	$95\frac{5}{8}$	$11\frac{3}{8}$
0.55	$88\frac{1}{2}$	$102\frac{5}{8}$	$14\frac{1}{8}$	$89\frac{3}{8}$	$101\frac{7}{8}$	$12\frac{1}{2}$	90	$101\frac{3}{8}$	$11\frac{3}{8}$
0.6	$94\frac{3}{8}$	$108\frac{1}{4}$	$13\frac{7}{8}$	$95\frac{1}{8}$	$107\frac{5}{8}$	$12\frac{1}{2}$	$95\frac{1}{4}$	107	$11\frac{1}{4}$
$0.625 = \frac{5}{8}$	$97\frac{1}{4}$	$111\frac{1}{8}$	$13\frac{7}{8}$	98	$110\frac{1}{2}$	$12\frac{1}{2}$	$98\frac{5}{8}$	$109\frac{3}{4}$	$11\frac{1}{8}$
0.65	$100\frac{1}{4}$	$113\frac{7}{8}$	$13\frac{5}{8}$	$101\frac{1}{8}$	$113\frac{1}{4}$	$12\frac{1}{8}$	$101\frac{3}{4}$	$112\frac{5}{8}$	$10\frac{7}{8}$
$0.666 = \frac{2}{3}$	$102\frac{3}{8}$	$115\frac{3}{4}$	$13\frac{3}{8}$	$103\frac{1}{8}$	$115\frac{1}{4}$	$12\frac{1}{8}$	$103\frac{3}{4}$	$114\frac{1}{2}$	$10\frac{7}{8}$
0.68	104	$117\frac{3}{8}$	$13\frac{3}{8}$	$104\frac{3}{4}$	$116\frac{3}{4}$	12	$105\frac{1}{2}$	$116\frac{1}{4}$	$10\frac{3}{4}$
0.7	$106\frac{5}{8}$	$119\frac{5}{8}$	13	$107\frac{3}{8}$	119	$11\frac{5}{8}$	108	$118\frac{1}{4}$	$10\frac{1}{4}$
0.71	$107\frac{7}{8}$	$120\frac{3}{4}$	$12\frac{7}{8}$	$108\frac{5}{8}$	$120\frac{1}{4}$	$11\frac{5}{8}$	109	$119\frac{1}{4}$	$10\frac{3}{8}$
0.73	$110\frac{1}{2}$	$123\frac{1}{4}$	$12\frac{3}{4}$	$111\frac{1}{4}$	$122\frac{5}{8}$	$11\frac{3}{8}$	112	$122\frac{1}{4}$	$10\frac{1}{8}$
$0.75 = \frac{3}{4}$	$113\frac{1}{4}$	$125\frac{5}{8}$	$12\frac{3}{8}$	114	$125\frac{1}{8}$	$11\frac{1}{8}$	$114\frac{5}{8}$	$124\frac{5}{8}$	10
0.76	$114\frac{5}{8}$	$126\frac{3}{4}$	$12\frac{1}{8}$	$115\frac{3}{8}$	$126\frac{3}{8}$	11	$116\frac{1}{8}$	$125\frac{7}{8}$	$9\frac{3}{4}$
0.77	$116\frac{1}{8}$	$128\frac{1}{8}$	12	$116\frac{7}{8}$	$127\frac{5}{8}$	$10\frac{3}{4}$	$117\frac{7}{8}$	$127\frac{1}{4}$	$9\frac{3}{8}$
0.78	$117\frac{5}{8}$	$129\frac{3}{8}$	$11\frac{3}{4}$	$118\frac{3}{8}$	$128\frac{7}{8}$	$10\frac{1}{2}$	119	$128\frac{1}{2}$	$9\frac{1}{2}$
0.79	$119\frac{1}{8}$	$130\frac{3}{4}$	$11\frac{5}{8}$	$119\frac{7}{8}$	$130\frac{1}{4}$	$10\frac{3}{8}$	$120\frac{1}{4}$	$129\frac{1}{2}$	$9\frac{1}{4}$
0.8	$120\frac{1}{2}$	132	$11\frac{1}{2}$	$121\frac{1}{4}$	$131\frac{1}{2}$	$10\frac{1}{4}$	$121\frac{7}{8}$	$131\frac{1}{8}$	$9\frac{1}{4}$
0.81	$122\frac{1}{8}$	$133\frac{1}{4}$	$11\frac{1}{8}$	$122\frac{3}{4}$	$132\frac{3}{4}$	10	$123\frac{1}{2}$	$132\frac{1}{2}$	9
0.82	$123\frac{5}{8}$	$134\frac{5}{8}$	11	$124\frac{1}{2}$	$134\frac{1}{4}$	$9\frac{3}{4}$	125	$133\frac{3}{4}$	$8\frac{3}{4}$
0.83	$125\frac{3}{8}$	136	$10\frac{5}{8}$	126	$135\frac{5}{8}$	$9\frac{5}{8}$	$126\frac{5}{8}$	$135\frac{1}{4}$	$8\frac{3}{4}$
0.84	127	$137\frac{1}{2}$	$10\frac{1}{2}$	$127\frac{5}{8}$	137	$9\frac{3}{8}$	$128\frac{1}{4}$	$136\frac{3}{4}$	$8\frac{1}{4}$
0.85	$128\frac{5}{8}$	$138\frac{7}{8}$	$10\frac{1}{4}$	$129\frac{3}{8}$	$138\frac{1}{2}$	$9\frac{1}{8}$	130	$138\frac{1}{4}$	$8\frac{1}{4}$
0.86	$130\frac{1}{2}$	$140\frac{3}{8}$	$9\frac{7}{8}$	$131\frac{1}{4}$	140	$8\frac{3}{4}$	$131\frac{1}{8}$	$139\frac{1}{4}$	$8\frac{1}{8}$
0.87	$132\frac{3}{8}$	142	$9\frac{5}{8}$	133	$141\frac{3}{4}$	$8\frac{3}{4}$	$133\frac{1}{2}$	$141\frac{1}{4}$	$7\frac{3}{4}$
$0.875 = \frac{7}{8}$	$133\frac{1}{4}$	$142\frac{5}{8}$	$9\frac{3}{8}$	$133\frac{3}{4}$	$142\frac{3}{8}$	$8\frac{5}{8}$	$134\frac{3}{8}$	$142\frac{1}{8}$	$7\frac{3}{4}$
0.88	$134\frac{1}{4}$	$143\frac{1}{2}$	$9\frac{1}{4}$	$134\frac{3}{4}$	$143\frac{1}{4}$	$8\frac{1}{2}$	135	$142\frac{7}{8}$	$7\frac{5}{8}$
0.89	$136\frac{1}{8}$	$145\frac{1}{8}$	9	$136\frac{3}{4}$	$144\frac{3}{4}$	8	$137\frac{1}{4}$	$144\frac{1}{4}$	$7\frac{1}{4}$
0.9	$138\frac{1}{8}$	$146\frac{3}{4}$	$8\frac{5}{8}$	$138\frac{5}{8}$	$146\frac{3}{8}$	$7\frac{3}{4}$	$139\frac{1}{4}$	$146\frac{1}{4}$	7
0.91	$140\frac{1}{2}$	$148\frac{1}{2}$	8	141	$148\frac{1}{4}$	$7\frac{1}{4}$	$141\frac{3}{8}$	148	$6\frac{5}{8}$
0.92	$142\frac{5}{8}$	$150\frac{3}{8}$	$7\frac{3}{4}$	$143\frac{1}{4}$	$150\frac{1}{8}$	$6\frac{7}{8}$	$143\frac{5}{8}$	$149\frac{7}{8}$	$6\frac{1}{4}$
0.95	$150\frac{5}{8}$	$156\frac{3}{4}$	$6\frac{1}{8}$	$151\frac{1}{8}$	$156\frac{1}{2}$	$5\frac{3}{8}$	$151\frac{1}{2}$	$156\frac{3}{4}$	$4\frac{7}{8}$

STROKE TABLE.

Piston Position. (Stroke=unity.)	CRANK ANGLES. (FOR ORDINARY CONNECTING ROD.)								
	RATIO 1 : 5½.			RATIO 1 : 6.			RATIO 1 : 6½.		
	Forward	Return.	Diff.	Forward	Return.	Diff.	Forward	Return.	Diff.
	deg.	deg.	deg.	deg.	deg.	deg.	deg.	deg.	deg.
0.125 = ⅛	38¼	45¼	7	38½	44¾	6¼	38¾	44½	5¾
0.2	49¼	57½	8¼	49½	57¼	7¾	49¾	56¾	7
0.25 = ¼	55¾	64¾	9	56¼	64⅜	8⅛	56⅜	64	7⅝
0.3	61⅞	71½	9⅝	62⅜	71	8⅝	62⅝	70⅝	8
0.333 = ⅓	65⅞	75⅝	9¾	66¼	75¼	9	66⅝	74¾	8⅛
0.375 = ⅜	70⅝	80¾	10⅛	71⅛	80¼	9⅛	71⅜	79⅞	8½
0.4	73½	83⅝	10⅛	73⅞	83¼	9⅜	74¼	82¾	8½
0.45	79¼	89½	10¼	79⅝	88⅞	9¼	80	88⅝	8⅝
0.5 = ½	84⅞	95⅛	10¼	85⅜	94⅝	9¼	85⅝	94⅜	8¾
0.55 ·	90½	100¾	10¼	91⅛	100⅜	9¼	91⅜	100	8⅝
0.6	96⅜	106½	10⅛	96¾	106⅛	9⅜	97¼	105¾	8½
0.625 = ⅝	99¼	109⅜	10⅛	99¾	108⅞	9⅛	100⅛	108⅝	8½
0.65	102⅜	112¼	9⅞	102¾	111¾	9	103⅛	111½	8⅜
0.666 = ⅔	104⅜	114⅛	9¾	104¾	113¾	9	105¼	113⅜	8⅛
0.68	106⅛	115¾	9⅝	106½	115⅜	8⅞	106⅞	115	8⅛
0.7	108½	118⅛	9⅝	109	117⅝	8⅝	109⅜	117⅜	8
0.71	109⅞	119¼	9⅜	110⅜	118⅞	8½	110¾	118⅝	7⅞
0.73	112½	121¾	9¼	113	121⅜	8⅜	113⅜	121⅛	7¾
0.75 = ¾	115¼	124¼	9	115⅝	123¾	8⅛	116	123⅜	7⅝
0.76	116⅝	125½	8⅞	117	125⅛	8⅛	117¾	124⅞	7½
0.77	118	126¾	8¾	118½	126½	8	118⅞	126¼	7¾
0.78	119½	128⅛	8⅝	120	127¾	7¾	120⅜	127½	7⅛
0.79	121	129⅜	8⅜	121⅜	129	7⅝	121¾	128¾	7
0.8	122½	130¾	8¼	122⅞	130½	7⅝	123¼	130¼	7
0.81	124	132	8	124⅜	131¾	7⅜	124⅝	131½	6⅞
0.82	125½	133½	8	126	133⅛	7⅛	126¼	132⅞	6⅝
0.83	127⅛	134¾	7⅝	127½	134½	7	127¾	134⅜	6⅝
0.84	128¾	136⅜	7⅝	129¼	136	6¾	129½	135¾	6¼
0.85	130⅜	137¾	7⅜	130¾	137½	6½	131¼	137⅜	6⅛
0.86	132¼	139⅜	7⅛	132½	139	6½	132¾	138⅞	6⅛
0.87	133⅞	141	7⅛	134¼	140⅝	6⅜	134½	140½	6
0.875 = ⅞	134¾	141¾	7	135¼	141½	6¼	135½	141¼	5¾
0.88	135¾	142⅝	6⅞	136¼	142⅜	6⅛	136½	142¼	5¾
0.89	137⅝	144¼	6⅝	138	144	6	138¼	143¾	5½
0.9	139¾	145⅞	6⅛	140	145⅝	5½	140	145¼	5¼
0.91	141¾	147¾	6	142⅜	147½	5⅜	142⅜	147⅝	5
0.92	144⅛	149⅝	5½	144⅜	149⅜	5	144½	149¼	4⅝
0.95	151¾	156⅛	4⅜	152	156	4	152¼	155¾	3½

STROKE TABLE.

Piston Position. (Stroke=unity.)	CRANK ANGLES. (FOR ORDINARY CONNECTING ROD.)								
	RATIO 1 : 7.			RATIO 1 : 7½.			RATIO 1 : 8.		
	Forward	Return.	Diff.	Forward	Return.	Diff.	Forward	Return.	Diff.
	deg.	deg.	deg.	deg.	deg.	deg.	deg.	deg.	deg.
$0.125=\tfrac18$	39	$44\tfrac14$	$5\tfrac14$	$39\tfrac18$	44	$4\tfrac78$	$39\tfrac14$	$43\tfrac78$	$4\tfrac58$
0.2	50	$56\tfrac12$	$6\tfrac12$	$50\tfrac14$	$56\tfrac38$	$6\tfrac18$	$50\tfrac38$	$56\tfrac18$	$5\tfrac34$
$0.25=\tfrac14$	$56\tfrac58$	$63\tfrac58$	7	$56\tfrac34$	$63\tfrac38$	$6\tfrac58$	$57\tfrac18$	$63\tfrac14$	$6\tfrac18$
0.3	$62\tfrac34$	$70\tfrac14$	$7\tfrac12$	$63\tfrac18$	70	$6\tfrac78$	$63\tfrac38$	$69\tfrac34$	$6\tfrac38$
$0.333=\tfrac13$	$66\tfrac78$	$74\tfrac12$	$7\tfrac58$	$67\tfrac18$	$74\tfrac14$	$7\tfrac18$	$67\tfrac38$	74	$6\tfrac58$
$0.375=\tfrac38$	$71\tfrac58$	$79\tfrac12$	$7\tfrac78$	$71\tfrac78$	$79\tfrac38$	$7\tfrac12$	$72\tfrac18$	$79\tfrac18$	7
0.4	$74\tfrac12$	$82\tfrac12$	8	$74\tfrac34$	$82\tfrac14$	$7\tfrac12$	75	82	7
0.45	$80\tfrac14$	$88\tfrac38$	$8\tfrac18$	$80\tfrac12$	88	$7\tfrac12$	$80\tfrac34$	$87\tfrac34$	7
$0.5=\tfrac12$	$85\tfrac78$	$94\tfrac18$	$8\tfrac14$	$86\tfrac14$	$93\tfrac34$	$7\tfrac12$	$86\tfrac38$	$93\tfrac58$	$7\tfrac14$
0.55	$91\tfrac58$	$99\tfrac34$	$8\tfrac18$	92	$99\tfrac12$	$7\tfrac12$	$92\tfrac14$	$99\tfrac14$	7
0.6	$97\tfrac12$	$105\tfrac12$	8	$97\tfrac34$	$105\tfrac14$	$7\tfrac12$	98	105	7
$0.625=\tfrac58$	$100\tfrac12$	$108\tfrac38$	$7\tfrac78$	$100\tfrac58$	$108\tfrac18$	$7\tfrac12$	$100\tfrac78$	$107\tfrac78$	7
0.65	$103\tfrac12$	$111\tfrac14$	$7\tfrac34$	$103\tfrac34$	$110\tfrac78$	$7\tfrac18$	104	$110\tfrac34$	$6\tfrac34$
$0.666=\tfrac23$	$105\tfrac12$	$113\tfrac18$	$7\tfrac58$	$105\tfrac34$	$112\tfrac78$	$7\tfrac18$	106	$112\tfrac58$	$6\tfrac58$
0.68	$107\tfrac14$	$114\tfrac34$	$7\tfrac12$	$107\tfrac12$	$114\tfrac12$	7	$107\tfrac34$	$114\tfrac14$	$6\tfrac12$
0.7	$109\tfrac34$	$117\tfrac14$	$7\tfrac12$	110	$116\tfrac78$	$6\tfrac78$	$110\tfrac14$	$116\tfrac58$	$6\tfrac38$
0.71	111	$118\tfrac38$	$7\tfrac38$	$111\tfrac14$	$118\tfrac18$	$6\tfrac78$	$111\tfrac12$	$117\tfrac78$	$6\tfrac38$
0.73	$113\tfrac58$	$120\tfrac78$	$7\tfrac14$	$113\tfrac78$	$120\tfrac58$	$6\tfrac34$	$114\tfrac18$	$120\tfrac38$	$6\tfrac14$
$0.75=\tfrac34$	$116\tfrac38$	$123\tfrac38$	7	$116\tfrac58$	$123\tfrac14$	$6\tfrac58$	$116\tfrac34$	$122\tfrac78$	$6\tfrac18$
0.76	$117\tfrac58$	$124\tfrac58$	7	$117\tfrac78$	$124\tfrac38$	$6\tfrac12$	$118\tfrac18$	$124\tfrac18$	6
0.77	$119\tfrac18$	126	$6\tfrac78$	$119\tfrac38$	$125\tfrac34$	$6\tfrac38$	$119\tfrac58$	$125\tfrac18$	$5\tfrac78$
0.78	$120\tfrac12$	$127\tfrac14$	$6\tfrac34$	$120\tfrac34$	127	$6\tfrac14$	121	$126\tfrac78$	$5\tfrac78$
0.79	122	$128\tfrac12$	$6\tfrac12$	$122\tfrac14$	$128\tfrac38$	$6\tfrac18$	$122\tfrac12$	$128\tfrac18$	$5\tfrac58$
0.8	$123\tfrac12$	130	$6\tfrac12$	$123\tfrac58$	$129\tfrac34$	$6\tfrac18$	$123\tfrac78$	$129\tfrac58$	$5\tfrac12$
0.81	125	$131\tfrac14$	$6\tfrac14$	$125\tfrac18$	$131\tfrac18$	6	$125\tfrac38$	$130\tfrac78$	$5\tfrac12$
0.82	$126\tfrac12$	$132\tfrac34$	$6\tfrac14$	$126\tfrac34$	$132\tfrac12$	$5\tfrac34$	127	$132\tfrac14$	$5\tfrac14$
0.83	$128\tfrac18$	$134\tfrac14$	$6\tfrac18$	$128\tfrac38$	$133\tfrac78$	$5\tfrac12$	$128\tfrac12$	$133\tfrac34$	$5\tfrac14$
0.84	$129\tfrac58$	$135\tfrac58$	6	130	$135\tfrac12$	$5\tfrac12$	$130\tfrac18$	$135\tfrac14$	$5\tfrac18$
0.85	$131\tfrac38$	$137\tfrac14$	$5\tfrac78$	$131\tfrac58$	137	$5\tfrac38$	$131\tfrac34$	$136\tfrac34$	5
0.86	$133\tfrac18$	$138\tfrac34$	$5\tfrac58$	$133\tfrac14$	$138\tfrac12$	$5\tfrac14$	$133\tfrac12$	$138\tfrac38$	$4\tfrac78$
0.87	$134\tfrac34$	$140\tfrac14$	$5\tfrac12$	135	$140\tfrac18$	$5\tfrac18$	$135\tfrac14$	140	$4\tfrac34$
$0.875=\tfrac78$	$135\tfrac34$	$141\tfrac18$	$5\tfrac38$	136	$140\tfrac78$	$4\tfrac78$	$136\tfrac18$	$140\tfrac34$	$4\tfrac58$
0.88	$136\tfrac38$	142	$5\tfrac38$	$136\tfrac78$	$141\tfrac34$	$4\tfrac78$	137	$141\tfrac58$	$4\tfrac58$
0.89	$138\tfrac12$	$143\tfrac58$	$5\tfrac18$	$138\tfrac34$	$143\tfrac12$	$4\tfrac34$	$138\tfrac78$	$143\tfrac38$	$4\tfrac12$
0.9	$140\tfrac12$	$145\tfrac38$	$4\tfrac78$	$140\tfrac34$	$145\tfrac14$	$4\tfrac12$	$140\tfrac34$	145	$4\tfrac14$
0.91	$142\tfrac58$	$147\tfrac14$	$4\tfrac58$	$142\tfrac34$	$147\tfrac18$	$4\tfrac38$	143	147	4
0.92	$144\tfrac78$	$149\tfrac18$	$4\tfrac14$	145	149	4	$145\tfrac18$	149	$3\tfrac78$
0.95	$152\tfrac38$	$155\tfrac58$	$3\tfrac14$	$152\tfrac12$	$155\tfrac58$	$3\tfrac18$	$152\tfrac58$	$155\tfrac12$	$2\tfrac78$

Referring to this Ratio column in the Stroke Table we obtain :—

Crank angle of the forward stroke for the 0.65 position $=112\frac{3}{8}°$.

Crank angle of the return stroke for the 0.65 position $=122\frac{1}{4}°$.

Difference between the return and forward$=9\frac{7}{8}°$.

SECOND EXAMPLE.

Stroke of piston$=45$ inches.

Ratio of crank to rod$=1:6\frac{1}{2}$.

Forward stroke crank angle$=131\frac{1}{4}°$.

Return stroke crank angle$=134\frac{3}{8}°$.

What locations will the piston occupy for these angles ?

From the Stroke Table we learn that :—$131\frac{1}{4}°$ forward$=$ piston location of 0.85 the stroke and $134\frac{3}{8}°$ return$=$piston location of 0.83 the stroke—consequently :

$45'' \times 0.85 = 38\frac{1}{4}$ inches from commencement of forward stroke.
$45'' \times 0.83 = 37\frac{3}{8}$ " " " return "

ECCENTRIC AND VALVE CONNECTION.

The principle of this connection has already been illustrated by Fig. 4, its standard motion in Fig. 5, but as the latter rarely occurs in practice it becomes necessary to study the former with reference to its influence on the events of the valve motion. It has been observed that the combination is nothing more nor less, than that of a small crank with a long connecting rod, the valve will therefore move in precisely the same manner as the piston, and will have in its progress from one extremity of the travel to the opposite, like irregularities, differing only in degree. In other words, when the eccentric arrives at the positions for cut-off and lead, the valve will be drawn *beyond* its true position—measured towards the eccentric—by a distance dependent on the ratio between the throw of the eccentric and the length of its rod. Since this difficulty is corrected by *lengthening* the rod, it follows that the *width* of the port opening in one stroke, will slightly *exceed* that in the other. This is practically the only effect produced by the use of the true eccentric connection ; although strictly speaking there is besides a slight difference in the equality of the exhaust closure, yet in no case does this become sufficient to affect the general action.

Neither is the difference in the opening appreciable in stationary engines, for their ratio of eccentric throw to length of rod is usually that of 1 : 20 or 30, which gives a variation too small to influence the general admission of the steam. (See Example 4 of Appendix

---

LENGTH OF ECCENTRIC ROD.

To find this dimension geometrically, after the proportions of the valve have been determined with the Travel Scale, proceed as follows :—

1st. Map the centre lines of the cylinder, valve, rocker, and main shaft.

2d. Place the crank at the zero and the valve in such a position that its edge F (Fig. 11) will give the proper opening for the *forward* stroke lead.

3d. Find the position of the centre of the eccentric by laying off the angular advance from the 90° line. Finally measure the distance from this centre to the rocker pin—or that of the valve stem as the case may be—for the correct length of the eccentric rod.

CORRECT CRANK ANGLE.

In finding the travel, lap, etc., of a valve appropriate to a given cut-off, by means of the Travel Scale and Stroke Table, the *Return Stroke crank angle should be invariably employed.* This will result in a correct cut-off for the return stroke although a *later* one than desired in the forward stroke. A method of correction for this inequality will presently be explained.

EQUALIZATION OF LEAD OPENING

Since the lead opening is measured a^t the moments when the crank pin passes the central line, there can exist no irregularity between the piston positions, and if the length of the eccentric rod be determined as above the width of the lead opening in the *return* stroke will *equal* that of the *forward* stroke.

EQUALIZATION OF EXHAUST CLOSURE.

The nearer we approach the half stroke positions of the piston the greater becomes the inequality between the crank angles, and as the uniformity of the exhaust closure and cut-off depends directly upon these values, it is evident

that they also will prove unequal for the forward and re
turn strokes.

The equalization of the exhaust closure is *always* possible, because for a *direct-action* engine a small amount of
inside lap added to the face N (Fig. 11) of the valve and
clearance taken from the face F, tends to *hasten* the moment of closure in one stroke and *delay* it in the other,
without producing the *slightest* effect on either the periods
of cut-off or the extent of the lead, both of which are con·
trolled solely by the outer edges of the vaive face ; *vice versa*
for a back-action engine.   Hence, any valve that is found
on examination to give unequal exhaust closure, clearly
proves a great lack of appreciation on the part of its designer, of those simple means which lie at his disposal for
the accomplishment of a perfectly correct exhaust and
release.

To illustrate this mode of equalization,

Suppose that in the General Example of Part I, the
crank and rod ratio was 1 : 4.

Its mean exhaust-closure angle being 142, or a piston
position of about 0.89 of the stroke, we have by the Stroke
Table a difference angle of about 9 degrees.

If then *one-half of this difference angle*, or $4\frac{1}{2}$ degrees,
be laid off from the Base line on the $5\frac{3}{8}''$ travel line of the
Scale (as in Fig. 12) we will obtain a correction of $\frac{1}{4}$ inch,
which added to the face N gives—

$$N = 2\frac{5}{16}'' + \frac{1}{4}'' = 2\frac{9}{16} \text{ inches, and subtracted,}$$
$$F = 2\frac{5}{16}'' - \frac{1}{4} = 2\frac{1}{4} \text{ inches.}$$

The true dimensions of the faces F and N, for imparting
equalized exhaust closure with the given ratio of 1 : 4.

EQUALIZATION OF THE CUT-OFF.

Where an attempt is made to equalize the cut-off, that of the exhaust closure should be disregarded, for the only process that it is possible to adopt in such a case would practically achieve both results.

Since the four periods, at which lead opening commences and cut-off occurs, are regulated by only two edges of one valve actuated by a single eccentric, it must be evident that where inequalities exist between any two of the four angular positions of the eccentric, their mutual dependence will be such that the periods for these can only be equalized at the expense of the other two.

Hence the equalization of the cut-off depends entirely on the expediency of giving a greater amount of lead to the one stroke of the piston than to the other stroke. With high-speed engines, such a change cannot be considered as desirable, but with those of low speed it may be made with very beneficial results. For with high speed the effect of lost motion can be most successfully overcome by a valve having equal lead in both strokes, but if the workmanship be perfect and the boxes carefully set up, the adjustment becomes applicable to all speeds.

To equalize the cut-off only two alterations are required in the valve motion, viz:

1st. *An increase of the angular advance.*
2d. *A lengthening of the eccentric rod.*

To illustrate the process, suppose we were required to equalize the cut-off and exhaust closure of an engine, differing only in respect to the connections, from that described in the General Example of Part I.

5

Let the ratio of crank to connecting rod$=1:5\frac{1}{2}$ and the forward stroke lead angle*$=4°$.

From the Stroke Table we find that the *return-stroke angle* for the above crank and rod ratio with $\frac{2}{3}$ cut-off$=114°$, or an angular advance for a valve without lead of $33°$. With a lead angle of $4°$ the trial angular advance becomes $=35°$ and the lap angle$=31°$. As the port opening equals $1\frac{1}{4}$ inches, we readily find from the Travel Scale the following terms:

Travel$=5\frac{1}{8}''$.   Lap$=1\frac{5}{16}''$, and forward lead$=\frac{3}{16}$ inch, nearly.

The difference between the forward and return-stroke crank angles for the above ratio and cut-off$=9\frac{3}{4}°$. *One-half of this difference* or $4\frac{7}{8}°$ forms the correction appropriate to the angular advance. Adding this quantity to the trial angular advance, we obtain $40°$ for the *true* angular advance of the eccentric. We thereby secure all the terms necessary for determining the length of the eccentric rod as already explained.

With an angular advance of $40°$ and no irregularity in the piston motion, the exhaust would close in *both strokes* at a crank angle of $140°$, or a piston position (Table A) of 0.883 the entire stroke$=26\frac{1}{2}$ inches. The lengthening of the eccentric rod, however, corrects the irregularity produced by the connecting rod and preserves this same point of exhaust closure to the motion.

By measuring the distance between the true and trial angular advances—$40°$ and $35°$—we obtain a distance $\frac{3}{16}$ inch, by which it is necessary to increase the minimum width of the bridge B, making its new value$=\frac{5}{8}+\frac{3}{16}=\frac{13}{16}$ inch.

* Since from the nature of the case, the return-stroke lead angle must greatly exceed that of the forward stroke, the latter quantity should be taken at only $3°$, $4°$ or $5°$ to prevent the former becoming excessive.

The lead angle of the return stroke will equal that of the forward stroke added to the difference between the cut-off angles of the crank, or $4° + 9\frac{3}{4}° = 13\frac{3}{4}$ degrees.

The lead opening of the return stroke, will equal twice the measured distance between the angular advances added to the forward lead, or $\frac{3}{16}'' \times 2 + \frac{3}{16}'' = \frac{9}{16}$ inch.

It should be observed that lengthening the eccentric rod increases the width of the return stroke port opening at the expense of that of the forward stroke, but so long as the width of the latter *exceeds* the minimum limit (page 23) for perfect admission, such inequality is actually a matter of not the *slightest* consequence.

To guard against the danger of a contracted opening, it will be advisable when finding the travel for a corrected motion to estimate the width of the port opening at least 1 *or* $1\frac{1}{4}$ *times* the width of the steam port.

EXPERIMENTAL TEST.

The greatest nicety and refinement of adjustment may be employed to produce an equal admission and exhaust of the steam throughout both strokes, and yet the attempt to equalize the power fail from lack of a judicious arrangement of the steam pipes and passages, on account of which the steam will enter one end of the cylinder with a lower average pressure than the other. This inequality occurs more frequently with vertical cylinders than horizontal ones, its extent can only be detected by the Engineer's most valuable friend, the Indicator, whose card ever furnishes clear and indubitable proof of the character, time and correlation of the various events taking place within the cylinder.*

To illustrate the character of this inequality, a few cards are here inserted which were taken from a shop engine, having a vertical unjacketed cylinder. The engine has been in service for several years. Its steam pipe, supported by the roof girders, descends to the centre of the cylinder's length where it enters passages for conducting the steam to the upper and lower ports. From the nature of the case the passage to the lower port is more direct than that to the upper, hence the steam encounters less opposition and enters with a higher average pressure than through the other. The *dimensions of the engine* are the following:

Diameter of cylinder$=10\frac{3}{16}''$; Stroke$=20\frac{5}{8}''$.
Connecting rod$=51\frac{1}{2}''$; Ratio 1 : 5.
Eccentric rod$=62\frac{1}{2}''$.
Travel of valve$=3\frac{5}{8}''$; Lead$=\frac{1}{8}$ inch.
Steam ports$=\frac{7}{8}'' \times 10\frac{1}{4}''$
Outside lap$=\frac{3}{4}''$; Inside$=\frac{3}{8}''$.
Average boiler pressure$=65$ lbs.; Velocity$=100$ Rev.

For the purpose of securing a well-defined card the boiler pressure was reduced to 42 lbs. per square inch, and the revolutions of the fly wheel—by means of a brake—to 40 per minute.

The result is given in Fig. 14, where the full line represents the card of the forward stroke, the dotted that of the back. In making comparisons of this kind, it should al-

* For a complete analysis of this instrument, its practical operation, etc., the reader is referred to Mr. Charles T. Porter's Treatise on the RICHARD'S STEAM INDICATOR, enlarged by F. W. Bacon, M. E., and published by D. Van Nostrand, New York.

ways be borne in mind, that the steam line of the one card must be combined with the exhaust of the other before the

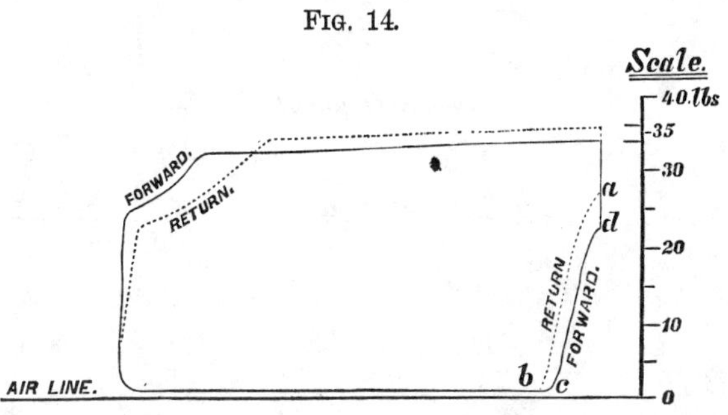

true relation is expressed between pressure and resistance. The difference in the steam pressure for the two strokes amounted to about 2 lbs. The cut-off took place at—

$$17\tfrac{1}{16} \text{ inches in the forward stroke}$$
$$15\tfrac{3}{4} \quad \text{``} \quad \text{``} \quad \text{return stroke.}$$

Difference$= 1\tfrac{5}{16}$ inches.

The compression commenced at $18\tfrac{1}{2}''$ forward.
    ``       ``   `` $17\tfrac{1}{2}$ return.

Difference$= 1$ inch.

This case is that of a valve motion in which no attempt has been made to equalize either the points of cut-off or exhaust closure. Had the exhaust closure been equalized the lines $a\,b$, $c\,d$, would have coincided.

Subsequently, the equalization of the cut-off and exhaust closure was accomplished with the same valve, by increasing its angular advance 5° and extending its eccentric rod.

The effect of these changes is shown in Fig. 15 ;* both

* The angles are constructed on the supposition that the piston is connected with the crank by means of a slotted cross-head and rod. The true angles of

Fig. 15.

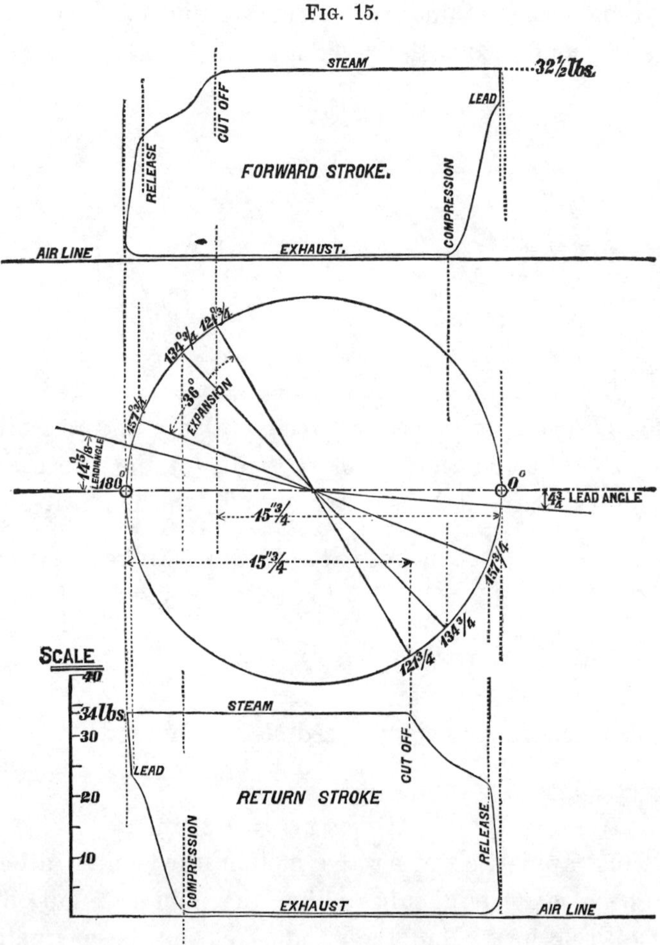

cut-offs took place at $15\frac{3}{4}$ inches, compressions at $17\frac{1}{2}$ (return stroke locations), the lead angle of the forward stroke became $4\frac{3}{4}°$ and that of the return $14\frac{5}{8}$, corresponding with an $\frac{1}{8}$ *inch* lead forward and an $\frac{1}{2}$ *inch* back.

the crank, however, for any ratio of crank to connecting rod, may readily be determined from these by first finding the piston positions due to these angles in Table A and then searching the Stroke Table for the forward and back stroke angles, appropriate to such positions.

It should be observed that this great inequality in the lead, so long as the port opening was ample, produced but trifling effect on the area enclosed by the card.

The two cards of Fig. 15 have been combined to form Fig. 16, showing an equality of the cut-off, exhaust closure

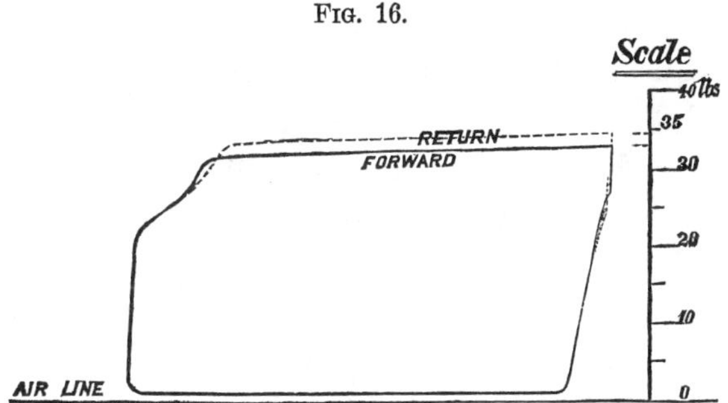

FIG. 16.

and release, but the same excess of pressure as before on the return stroke. Had the cylinder lain in a horizontal position, the latter inequality would have entirely disappeared, and equalized power been applied to both strokes of the piston.

HOW TO SET A SLIDE VALVE

HAVING EQUALIZED EXHAUST.

1st. Place the crank at the 180° location, mark on the cross-head and one of its guides opposing " centre punch " points.

2d. Bring the crank to the zero and mark a second point on the guide. The two points thus found, measure the length of the stroke. Move the eccentric until the valve has the required lead for the forward stroke.

3d. Advance the crank in the direction of the motion

until the exhaust of the opposite stroke closes ; scribe a line across the guide which shall pass through the point on the cross-head.

4th. Move the crank until the other exhaust closes and scribe a second line on the guide.

5th. If now the exhaust should close at equal distances from the commencement of each stroke the motion would be in adjustment ; if not, alter the length of the eccentric rod until the closure becomes equalized, then return the crank to the zero position, and alter the angular advance of the eccentric until the required lead of the forward stroke is secured.

The position of the valve at the moment of closure may readily be fixed by means of a " valve gauge" fitting centre punch points on the valve stem and its stuffing box.

The above process will serve also to equalize the cut-off if the valve be proportioned for this object.

PART III.

ADJUSTABLE ECCENTRICS.

ADJUSTABLE ECCENTRICS.

THE intensity of the resistances, opposed to the action of steam upon the piston of an engine, fluctuates in a very marked manner, and were it possible to meet variations instantaneously by an increase or diminution of the power applied, the result would be an impulsive action of the working parts instead of a smooth continuous motion.

It is obvious then that the power of an engine must be exerted at different periods of time in overcoming a more and more determined resistance or even one acting in an opposite direction.

Slight variations in the resistances are controlled or absorbed by the momentum of the working parts, as by the fly wheel of a stationary engine, the drivers and coupled wheels of a locomotive, and the screw or paddle wheel of a steamer.

But variations in degree must be met by an increase or diminution of the power applied. This may be accomplished in two ways, either by controlling the pressure of the steam *before* it enters the cylinder or by limiting it *after* an entrance has been effected, that is by cutting off at an earlier or later point of the piston's stroke. The first method deals with the throttle valve, the second relates to the closing of the steam port by its valve, while for marine purposes both of these agencies are usually employed.

The throttle may be operated either by hand or by an automatic device called a "governor," the principle of which has been so clearly elucidated in almost every work on the Steam Engine that it would appear superfluous to repeat it in this connection. The governor is also at times used as a regulator of the cut-off, a work it most beautifully and delicately executes in such stationary engines as the "Corliss," "Allan," and others. In these the steam and exhaust ports are perfectly distinct, so that a change in the cut-off does not affect the action of the exhaust, which remains from the beginning to the end of the stroke free of the back pressure peculiar to a single eccentric motion.

The principal regulator of the cut-off in modern practice, however, is the "link," or combined double eccentric motion, which retains, while presenting itself in various kindred forms, the individuality of the double eccentric. On this account the study of the action of an *adjustable* single eccentric forms the best introduction to a complete understanding of the link motion.

Hitherto we have considered the eccentric as a fixed body keyed to the main shaft, with its centre line or throw inclined at a certain angle to the crank arm. Thus in Fig. 17 the eccentric has an angular advance of $33\frac{1}{2}°$ or a position C F, suitable for a positive or forward motion of the crank if a rocker intervenes between the eccentric and the valve. It will be remembered (Fig. 9) that the same eccentric, when located with an opposite angular advance C B, holds a position appropriate to a negative or back motion of the crank. Hence if it be required to change the direction of the engine's motion, it is only necessary to unkey the eccentric, slip it around the shaft and secure its throw in a position C B, opposite to that of C F.

But if in addition to a change of direction a variation in the cut-off for the forward and back motions of the crank were demanded, subject to the single condition that the lead opening and lap shall remain unaltered, then it would be very clear that the eccentric must be slotted in such a manner that its centre can pass in a *direct* line from F to B (Fig. 17) instead of following the arc F *m* B. By this

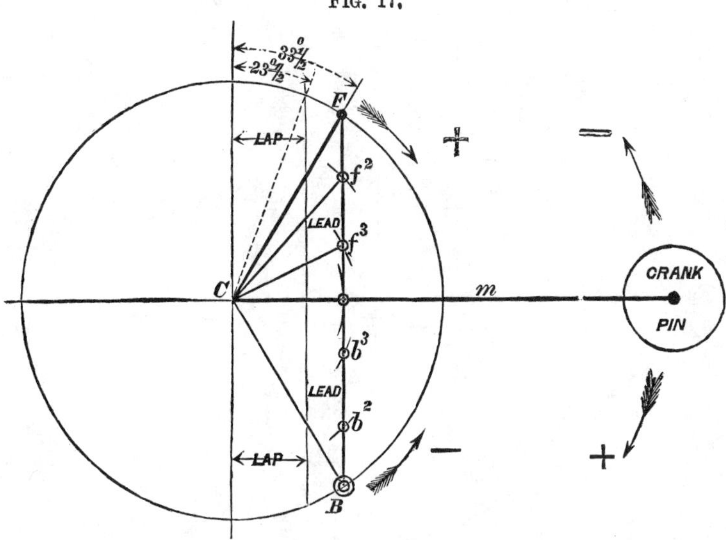

FIG. 17.

means the throw of the eccentric and consequently the travel of the valve would be reduced between F, the full gear forward of the eccentric, and B the full gear back, while the minumum value would lie at the point M midway between these extremes, or at the mid gear of the eccentric. When the eccentric is placed at f^2, the $\frac{1}{2}$ travel of the valve equals $c\,f^2$; at $f^3 = c f^3$ and at the mid gear $= $ C M. It need scarcely be told one already familiar with the use of the Travel Scale, that by thus reducing the travel of the valve—while the lap and lead opening are preserved constant—we obtain a reduced port opening, an increased angular advance, and consequently an earlier cut-off. Before proceeding, however, to analyze the effects of these changes one other case should be considered, namely that in which the lead opening admits of slight variation. The most common form in which this occurs is for the lead opening to *increase* from the full gears towards the mid gear as in Fig. 18, where the centre of the eccentric travels in an arc.

FIG. 18.

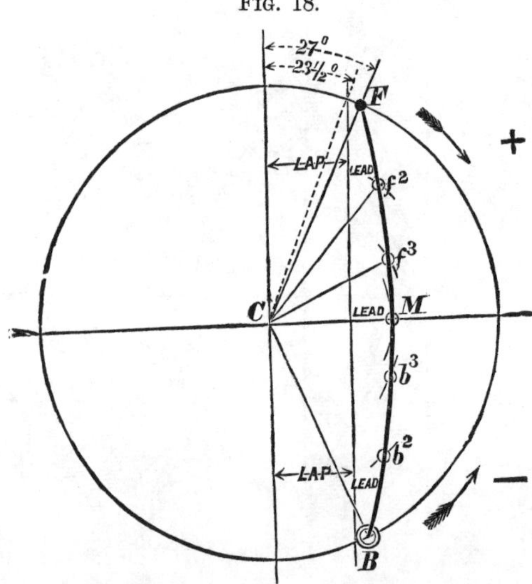

The reverse, or case of *diminution*, is so similar in principle, that the examination of one will serve to indicate the nature of the other.

Suppose, FOR EXAMPLE, that C F, the full gear throw of the eccentric.................................... $=2\frac{1}{2}$ inches.

Then the greatest travel of valve............... $=5$ "

Let the lap .... $=1$ "

And lead opening.............................. $= \frac{3}{8}$ "

These dimensions are chosen on account of no peculiar fitness in their relation to one another, but merely to illustrate the effect of reduced travel, constant lap, and a variable or constant lead upon the cut-off, lead angle, port opening and exhaust closure of a simple slide valve.

With a *Constant Lead opening* the centre of the eccentric should be moved in a direct line from F to B, Fig. 17. When located at F, the travel of the valve equals twice C F=5 inches, the lap=$1''$, and lead=$\frac{3}{8}''$; if now we mark the lap and lead on a slip of paper and apply it in the usual manner to the 5 inches line of the Travel Scale, we find that

The cut-off=$123°$=0.77 stroke.
The lead angle=10 degrees.
The port opening=$1\frac{1}{2}$ inches.
The exhaust closure=$146\frac{1}{2}°$=0.92 stroke.

Upon removing the centre of the eccentric from F to f' the travel will be reduced to twice C F^2=$4''$. Lap and lead opening remaining unchanged, we find by applying them to the $4''$ line of the Travel Scale that—

The cut-off=$106°$=0.64 stroke.
The lead angle=14 degrees.
The port opening=1 inch.
The exhaust closure=$136°$=0.86 stroke.

Since b^2 is removed the same distance from B that f is from F, the two positions f^2 and b^2 will give precisely the same results for opposite motions of the crank.

In like manner the cut-off, etc., for the positions f^3, b^3, and M may be readily found, but their mutual relation can best be exhibited by grouping their results in a Table like the following for—

Constant Lead.

Eccentric location.	Travel.	Lap.	Lead opening.	Cut-off.		Lead angle.	Port opening.	Angular advance.	Exhaust Closure.	
	deg.	deg.	ins.	deg.		deg.	deg.	deg.	deg.	
F or B	5	1	$\frac{3}{8}$	123	$=0.77$ stroke.	10	$1\frac{1}{2}$	$33\frac{1}{2}$	$146\frac{1}{2}=0.92$	stroke.
f^2 or b^2	4	1	$\frac{3}{8}$	106	$=0.64$ "	14	1	44	$136 =0.86$	"
f^3 or b^3	3	1	$\frac{3}{8}$	71	$=\frac{1}{3}$ "	25	$\frac{1}{2}$	67	$113 =0.695$	"
M	$2\frac{3}{4}$	1	$\frac{3}{8}$	$43\frac{1}{2}$	$=0.14$ "	$43\frac{1}{2}$	$\frac{3}{8}$	90	$90 =\frac{1}{2}$	"

Again, with *Variable Lead opening*, diminishing from the mid gear toward the full gears (from $\frac{3}{8}''$ to $\frac{1}{8}''$) the centre of the eccentric must move in the arc of a circle, as shown in Fig. 18, where the points f^2, f^3 are drawn a little nearer toward the centre of the shaft, thereby reducing the travel of the valve. A Table similar to the above has been constructed for showing the effect on the motion of—

Variable Lead.

Eccentric locations.	Travel.	Lap.	Lead opening.	Cut-off.		Lead angle.	Port opening.	Angular advance.	Exhaust Closure.	
	deg.	deg.	ins.	deg.		deg.	deg.	deg.	deg.	
F or B	5	1	$\frac{1}{8}$	$129\frac{1}{2}$	$=0.82$ stroke.	$3\frac{1}{2}$	$1\frac{1}{2}$	27	$153 =0.945$	stroke.
f^2 or b^2	$3\frac{7}{8}$	1	$\frac{1}{4}+\frac{1}{32}$	107	$=0.65$ "	$10\frac{1}{2}$	$1\frac{5}{16}$	$41\frac{1}{4}$	$138\frac{1}{2}=\frac{7}{8}$	"
f^3 or b^3	$2\frac{15}{16}$	1	$\frac{5}{16}+\frac{1}{32}$	71	$=\frac{1}{3}$ "	24	$\frac{7}{16}+\frac{1}{32}$	$66\frac{1}{2}$	$113\frac{1}{2}=0.7$	"
M	$2\frac{3}{4}$	1	$\frac{3}{8}$	$43\frac{1}{2}$	$=0.14$ "	$43\frac{1}{2}$	$\frac{3}{8}$	90	$90 =\frac{1}{2}$	"

From a comparison of the foregoing tables we are enabled to draw the following general conclusions:

I. That whether the eccentric centre be moved along a straight line, or upon an arc, the cut-off and exhaust closure will take place the earlier, the nearer its approach to the mid gear; while the lead angle and angular advance of the eccentric will be increased in magnitude.

II. At the mid gear—

The port opening=the lead opening of the valve.

Cut-off angle=lead angle of the valve.

Angular advance=90°.

Hence the exhaust closes at the ½ stroke.

III. If the port opening=0, in the mid gear, then the lead opening=0, and the lead angle=90°. The steam consequently is not admitted to the cylinder for either stroke of the piston.

IV. Since the free admission of the steam depends directly on the area of the port opening, and this is gradually reduced towards the mid gear, it must be evident that the *most influential* element upon the perfect action of a link motion, or adjustable eccentric, will be this width of lead opening in the mid gear; at which location it invariably equals the port opening.

V. Whether the lead opening remains constant or varies from the full to the mid gear, the lead angle—or angular distance of the crank from the nearest centre at the moment the port commences to open for the admission of steam—*increases* in magnitude from the full to the mid gear. At this point, if the lead opening=0, the angular distance becomes=90°. This condition is of course incompatible with motion, for the port opening being 0, no steam can enter to urge forward the piston, but could this be moved to the 90° or ½-stroke point by extraneous force, it would then be repelled by the steam admitted for the supply of the return stroke.

VI. To estimate, therefore, the effect of any lead on the continuity of the crank motion, it is necessary to know not only the width of the lead opening (or lead) at the end of the stroke, but also the angular distance of the crank (or lead angle) at the moment lead commenced. In other

6

words, we must know the *area* of the opening and the *time* required to produce it. Thus with a constant lead of $\frac{3}{8}$ inch the port commenced opening for the full gear when the crank was within 10° of the end of the stroke, but for the mid gear when the crank was within $43\frac{1}{2}$° of the same position, a portion of time nearly $4\frac{1}{2}$ times greater for the same opening. Since the volume of steam admitted through a contracted passage, depends more directly upon the time that passage is kept open than upon its area, it manifestly would be absurd to imagine that a valve motion, having *constant* lead, was capable of admitting steam with invariable effect for all points between the full and mid gears. So far as preadmission alone is concerned, the lead opening should grow *smaller* on the approach towards the mid gear.

The relations between travel, lap, lead and port opening for the different positions F, f^2, f^3, M, b^3, b^2, B of the eccentric, are conveniently expressed by a single diagram like Fig. 19.

Describe the travel circle a' F B, lay off on both sides of the centre or exhaust-closure line c' C, the lap and lead, and draw the cut-off and lead lines. Through the various positions F, f^2, etc., draw indefinite lines parallel to the central line of motion a' d', and upon these lay off on each side of the exhaust-closure line the $\frac{1}{2}$ travel for the location in question. Thus, from c' lay off c' d and c' a each equal to C F the $\frac{1}{2}$ travel; from c^2, c^2 h and c^2 e each equal to Cf^2, and so on. Finally, draw the curved lines d M d^2, a M^2 a^2 through the extremities of the travels for their boundary lines,* and we obtain a sort of travel scale expressive

---

* The resulting curves will invariably constitute the branches of an equilat eral hyperbola.

of the character of the motion at the different gears of the eccentric.

FIG. 19.

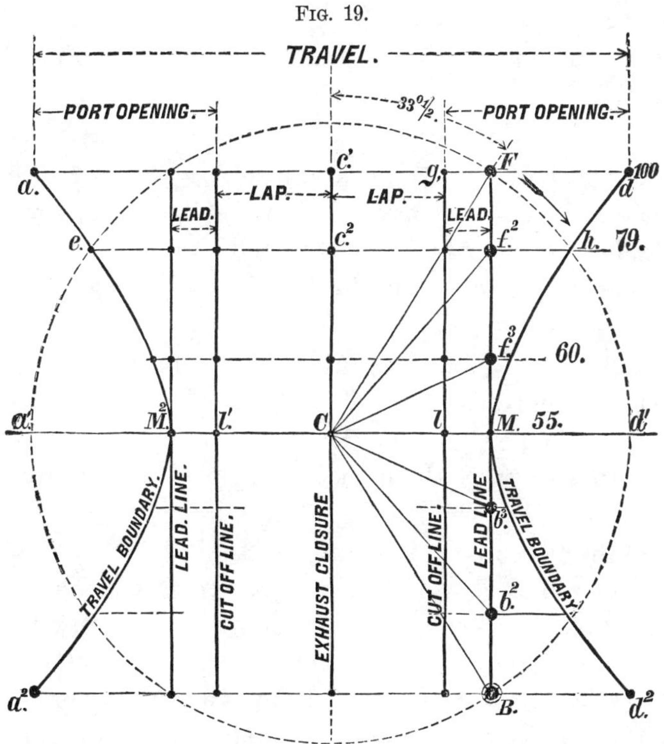

Since for any particular case the lap absorbs a constant portion of the travel, as the distance $l\,l'$, it is evident that in order to obtain a variable cut-off, we have sacrificed port opening and consequently the travel in both strokes, and have reduced its excess F d or B d^2 over the lead in the full gears, to *nothing* in the mid gear. This Figure will enable us hereafter to trace the analogy between link and adjustable eccentric motions.

Having explained the effects attending a motion of the eccentric's centre across its shaft, it is proper to indicate how such motion can be secured.

The devices are naturally grouped under two classes:

1st. Those capable of acting upon the eccentric while the engine is in motion.

2d. Those that require a state of rest for their manipulation.

The idea of moving the centre of an eccentric directly across its shaft, by means of two wedges with reversed points, first originated with Mr. Dodd, of Newcastle, England, and a device similar to that shown in Figure 20 was patented by him in 1839. His object was simply to obtain a reversable valve motion, little dreaming at the time of its latent powers as a variable cut-off. These were subsequently developed by Mr. Dubs of Glasgow and successfully applied, with slight modifications of design, to many locomotives.

In the accompanying figure the crank A is placed at the zero, as well as at right angles to its mate G. The valve motion of its cylinder is operated by the eccentric pulley D, whose centre f^2 may be moved across the shaft to any other position between F and B, by withdrawing or inserting the wedges $d\ d$ secured on one side of the sliding pulley H. In its opposite side are two similar wedges $e\ e$, for operating the pulley E, which controls the valve of the cylinder belonging to the crank G, and consequently moves in a line $n\ n^2$ at right angles to F B. If now the pulley H be slid along the crank shaft until brought in contact with the eccentric D, the centre f^2 will be transferred from f^2 to B, and in like manner n^l to n^2, locations appropriate to a back or negative motion of the crank. When however it comes in contact with the eccentric E, the two eccentrics will stand in the full gear forward positions F and n, and the nearer the pulley H approaches the mid position between

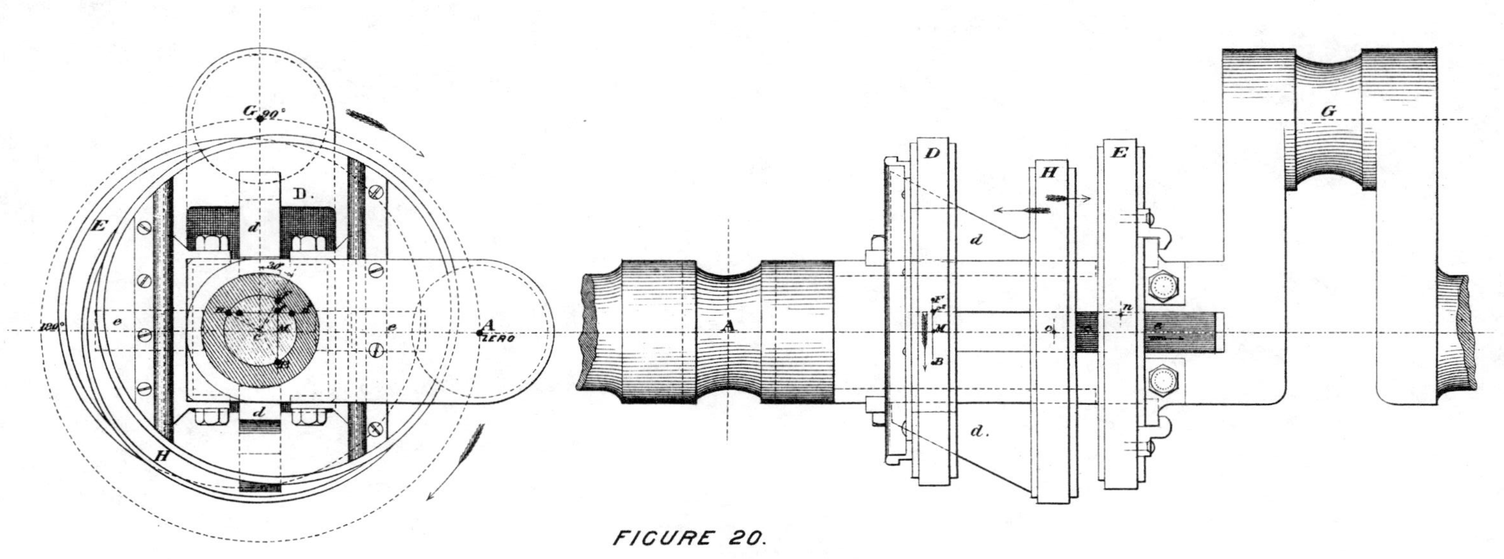

FIGURE 20.

D and E the earlier will the cut-off take place, as already shown in Fig. 17.

A moment's reflection will satisfy the mind, that the Dodd Wedge Motion cannot be corrected for inequalities of cut-off and exhaust closure due to angularity of the connecting rod, without introducing mid gear inequalities of port opening, far more pernicious in their effects than the evils we seek to abate.

The second class of motors find their chief application in connection with portable engines where it is seldom necessary to reverse the direction of the crank motion. The mechanical arrangement of the parts is shown in Fig. 21.

FIG. 21.

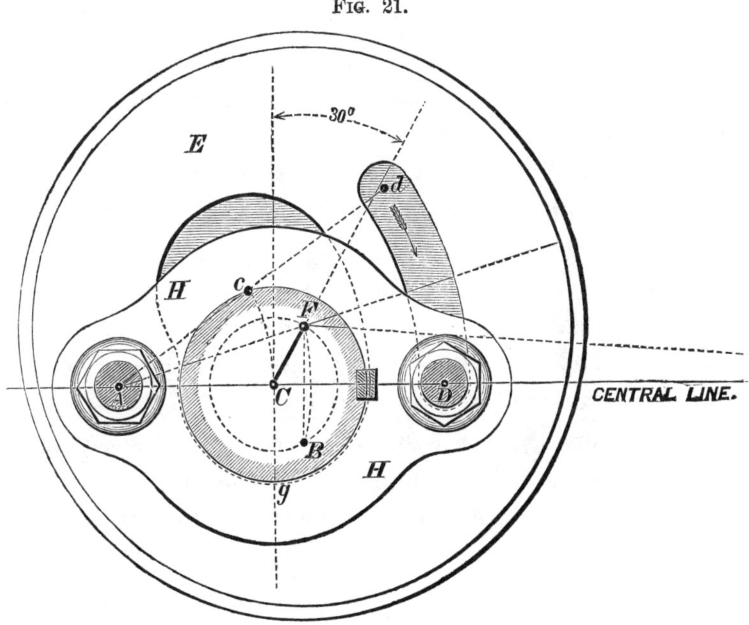

The collar H, being firmly keyed upon the main shaft $c\,g$, carries on one side a bolt A which performs the office of a pivot to the slotted eccentric pulley E. Through the opposite side passes the bolt D fitting loosely the slot D d, and capable of clamping the pulley in any desired position. The eccentric has an angular advance of 30° and its centre occupies the full gear forward position F.   To place it in the full gear back position at B, the bolt D should be loosened and the pulley E turned on its pivot A until the point d assumes the position D.   This motion will transfer the centre F, upon an arc whose radius equals A F, across the shaft to B, and give as in Fig. 18 an increasing lead from the full to the mid gear.   If however it were required that the lead be preserved constant for all gears of the eccentric, the curvature of the slot D d should be formed with a shorter radius than D A, a similar change should be made in the large slot, and the hole for the bolt A should be oval in the direction of the line A F so that the pin D might force the centre of the eccentric to adhere practically to the straight line joining F and B.

Figure 21 illustrates a simple method for determining the diameter of the eccentric pulley and for arranging the centres in such a manner that the large slot shall be disposed symmetrically on opposite sides of the diameter A F.   The process is carried out as follows :

1st. About the centre C describe a circle c g, with a diameter equal to that of the proposed shaft.

2d. Decide upon a suitable diameter for the eccentric pivot A and locate the bolt as near to the shaft c g, on the central line of motion, as consistent with the strength of the pulley.

3d. With A as a centre and A C radius describe an arc C c; join A with its point of intersection c of the shaft circle.

Bisect the angle C A c by the line A F.

4th. Lay off the required angular advance line C F.   Its intersection with the line A F locates the centre of the eccentric in the full gear forward.   About F as a centre strike a circle for the eccentric enclosing the pin A by a distance sufficient to give ample strength to the eccentric pulley. The distance C F will then equal the $\frac{1}{2}$ travel of the valve, with which radius the travel circle F B may be described, and the lap readily determined.

The angular advance is usually taken at about 30° or the $\frac{3}{4}$ cut-off, but of course the larger the advance the greater will be the travel of the valve, lap, &c

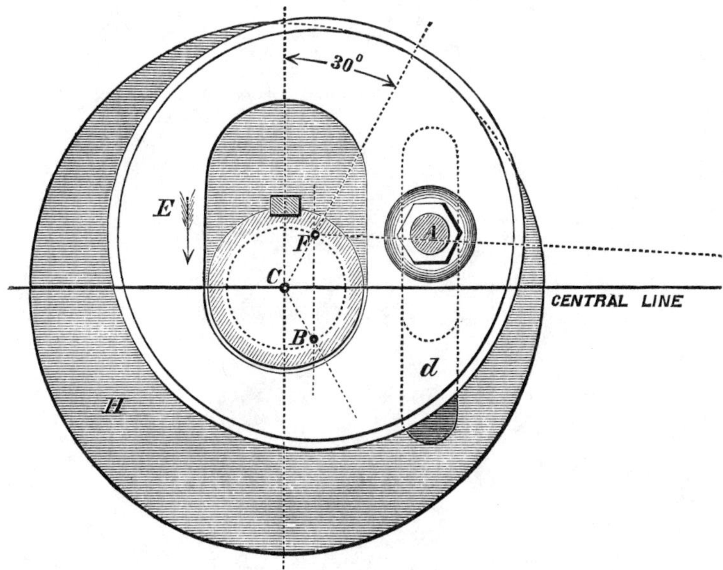

The above cut illustrates another modification of the adjustable eccentric.   In this device, the collar H is made larger than the eccentric pulley, and keyed fast to the shaft.   The pulley E is secured by the bolt A, and any tendency to its rotation prevented by means of a feather on the back, which fits the slot d.

PART IV.

LINK MOTIONS.

LINK MOTION.

The various mechanical devices embraced under this general term, have many strong points of resemblance and subserve a common object. By means of them, the Engineer is able at will to change the direction of the crank rotation, with only the loss of the time required for overcoming the momentum of the moving parts, and developing the like in a reverse direction. More than this simple result was not contemplated in the original discovery of the link. Subsequently, however, it was found to be capable of regulating the cut-off of the steam, so that the power could always be adjusted to the work required. This feature greatly enhanced its value, and placed the engine under the complete control of the operator.

The extreme simplicity of the parts of the link motion, has enabled it to contend successfully with all rivals, and at the present day it remains in substantially its primitive form. It is applied principally to locomotive and marine engines, where the power demanded is quite variable, and the motion at one time direct, at another reverse.

The designs may be divided into four classes:

I. The shifting link motion.
II. The stationary link motion.
III. The Allan link motion.
IV. The Walschäert link motion.

The first form was invented by Mr. Howe, in 1843, and applied to the locomotives of Messrs. Robert Stephenson & Co.   It is in fact the representative link motion, which, excepting slight modifications in the mode of suspension, remains unchanged by the accumulated experience of a quarter of a century.

Simultaneous with the appearance of this motion was that of the second, the discovery of Mr. Daniel Gooch.   It accomplishes perfectly analogous results, and has met with much favor throughout Great Britain and the Continent.

The "Allan" combines the characteristic features of the Howe and Gooch link motions in such a manner that the parts are more perfectly balanced, consequently it dispenses with the counter weight or spring peculiar to the former of these motions.

The Walschäert motion is extensively applied in Belgium, but probably will not receive much attention from locomotive Engineers, beyond the limits of that Kingdom, unless future designers succeed in reducing the number of its connections.

It is proposed to confine our investigation to the shifting link motion, to develop the general laws governing its action amid varied conditions, to present graphic methods for determining the proportions of the parts, and briefly to point out the general application of the same to the link motions of the other three Classes.

SHIFTING LINK MOTIONS.

A link, operated by two fixed eccentrics, forms when properly suspended an exact mechanical equivalent of the movable eccentric.   Unlike the latter, however, its motion is

capable of an accurate adjustment, which practically nulli-
fies the effect of irregularities in cut-off and exhaust closure,
attributable to the angularity of the main connecting rod.

The general form in which its parts are arranged in
American locomotive practice, is clearly shown in Fig. 22.
Upon the main shaft are keyed the forward and backing
eccentrics, with their centres at F and B, the two extreme
positions of the single movable eccentric in Fig. 17. Their
straps are bolted to the eccentric rods, and these in turn
are pinned to the "link." The slide valve is attached by
its stem to one of the rocker arms, and a "block" sur-
rounds the pin of the opposite arm, which fits the main
link and slides freely therein. The centre of the link is
spanned by a plate called the "saddle," on which is formed
the pin or stud that supports the link and eccentric rods.
This pin is embraced by a bar called the "hanger," or
sometimes the suspending or the sustaining link, from its
position and the service rendered to the motion. The
former term is preferable on account of its conciseness, and
can lead to no confusion. The opposite extremity of the
hanger is attached to one arm of the tumbling shaft. Both
arms of this shaft are rigidly secured, and form upon it a
"bell crank." The shaft itself freely oscillates on properly supported bearings, but is limited in its motion by the
action of the reversing rod. The link has been dropped
into the full gear forward, thus throwing the entire influence of the eccentric F upon the valve motion to the almost
complete exclusion of that of its mate B. By drawing back
the reversing rod and raising the link until the pin of the
other eccentric rod is brought in line with the pin of the
rocker arm, the link will be made to occupy a location appropriate to a negative crank movement (as with B, Fig. 17)
and *intermediate suspensions* will in like manner be pro-

ductive of *earlier* cut-off and exhaust closures. In order to clearly demonstrate that such similarity exists between these motions, it will be necessary to reduce Fig. 22 to a skeleton form like Fig. 23, and follow the journeyings of the "link arc" throughout a complete revolution of the crank.

Let the path of the main crank pin be represented by the circle E D in Fig. 23. This being divided into 12 equal parts, gives a sufficient number of positions for the purpose of tracing the motions of the link arc. The zero will be known as position No. 1, the 180° as position No. 7, and so on. Within this circle describe the path of the eccentric centres by means of the circle F B b^{7}. This should first be divided into 12 equal parts, with F as the origin of one eccentric's motion, and again into 12 other equal parts with B as an origin, so that when the crank moves from position No. 1 to 3 the new positions f^{3} and b^{3} of the two eccentrics may be instantly found, and the same with other locations. The original positions F and B are of course laid off with the angular advance due to the proposed maximum cut-off. At the distance C t from the centre of the shaft erect the perpendicular T t and locate T the fixed centre of the tumbling shaft. T h will represent the arm which supports the link through its hanger and h h' h'' the arc described by this arm. A second perpendicular at the distance C A will contain the point R, the centre of the rocker shaft, whose arm R A sweeps the arc r A r. The motion of the upper arm, being merely the reverse of the lower, need not be considered, and so long as the angular advance is properly located no error can arise from the omission. In the motion of the lower arm there are five locations of vital importance, viz: one at which the exhaust of the valve opens or closes, two appropriate to the lead at full gear of the link, and two at which cut-off takes place or the valve closes its ports. The 1st is evidently the normal position R A of the rocker, the 2d R d, R d', that in which the rocker pin is drawn aside a distance A d equal to the sum of the lap and lead, and the 3d R l, R l' corresponds with a removal A l equal to the lap. Hence, so far as the slide valve is concerned we can confine our attention to the motion of the rocker arm pin upon the arc r r. The five positions in question can be distinctly located by sweeping a circle d d', with a radius equal to the lap plus the lead of the valve, around the exhaust point A, and inscribing a second circle l l' with a radius equal to the lap of the valve. Then the four points in which these circles intersect the arc r r will give the 4 positions of the pin corresponding with the lead and cut-off positions of the valve, and the centre of these circles will give the exhaust closure positions. As these locations will be constantly referred to in the sequel, it should be remembered that the "lead circle" d d' fixes those points on the arc r r which the pin of the rocker arm must occupy when the valve has a given lead; and that the "lap circle" l l' locates the positions of the same pin for the moments at which the steam ports are closed against the admission of steam to the cylinder.

Our next duty will be to reduce the link to its simplest form.

It appears on examination that the rocker pin is entirely subject, in its motion, to the guidance of the link arc, and that this arc swept with a radius C A is rigidly connected with three moving points, viz. the saddle pin, and the two eccentric rod pins. In following the motion of the link arc, the connection of the parts can best be maintained by the use of a template, cut from white holly veneer or other hard wood and shaped like L L in Fig. 23, upon which are made $\vee$ shaped incisions for locating the points f, S and b of the pins.

We are now prepared to find position No. 1 of the link corresponding with No. 1 of the crank. Of course when the crank is at the zero the steam port should be opened an amount equal to the lead of the valve. The rocker arm therefore will occupy the position R d, and the point d lie in the link arc. Since the eccentric centres F, B are found in a line perpendicular to the central line of motion, and the eccentric rods are of equal length, the link must occupy a nearly perpendicular position. Place the template so that its arc coincides with the point d and mark the point f upon the paper, then the distance from F to f will equal the length of the eccentric rod. With this length as a radius describe about F as a centre the small arc $f\,g$, likewise with B as a centre describe the small arc $b\,h$. Apply the template to these arcs so that the points f and b shall be found in them and the point d on the link arc $n\,c\,d$, after which, draw the link arc on the paper and we obtain position No. 1 of the link. With the saddle pin S as a centre and the length of the hanger h S as a radius, the position T h of the tumbling shaft arm is readily found for full gear of the link and conversely the arc c S c is fixed along which the saddle pin must travel during the revolution of the crank.

The preparatory stages of our solution are now complete, the link motion of Fig. 22 has been reduced to its skeleton form and the first position of the link located. Our next step is to follow the link arc during its journeyings in a single revolution of the crank. Suppose then, the crank is made to occupy position No. 2, the eccentrics will be carried forward from F and B to $f^2\,b^2$. Since the length of their rods remains unchanged the arcs $f\,g$, $b\,h$, will be

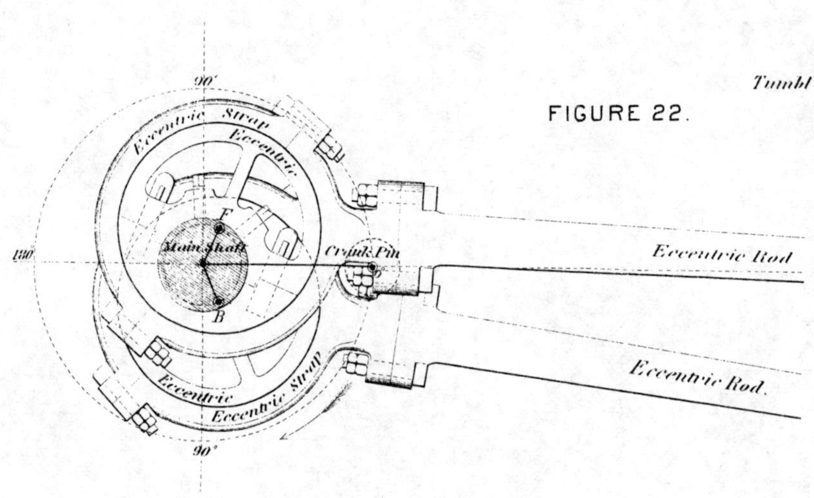

FIGURE 22.

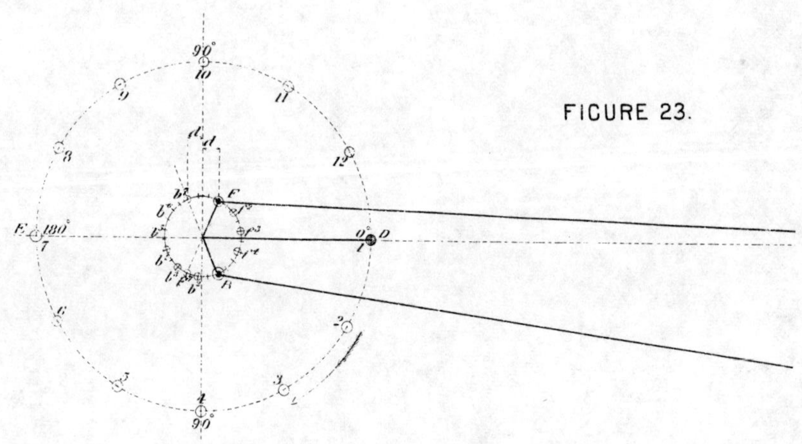

FIGURE 23.

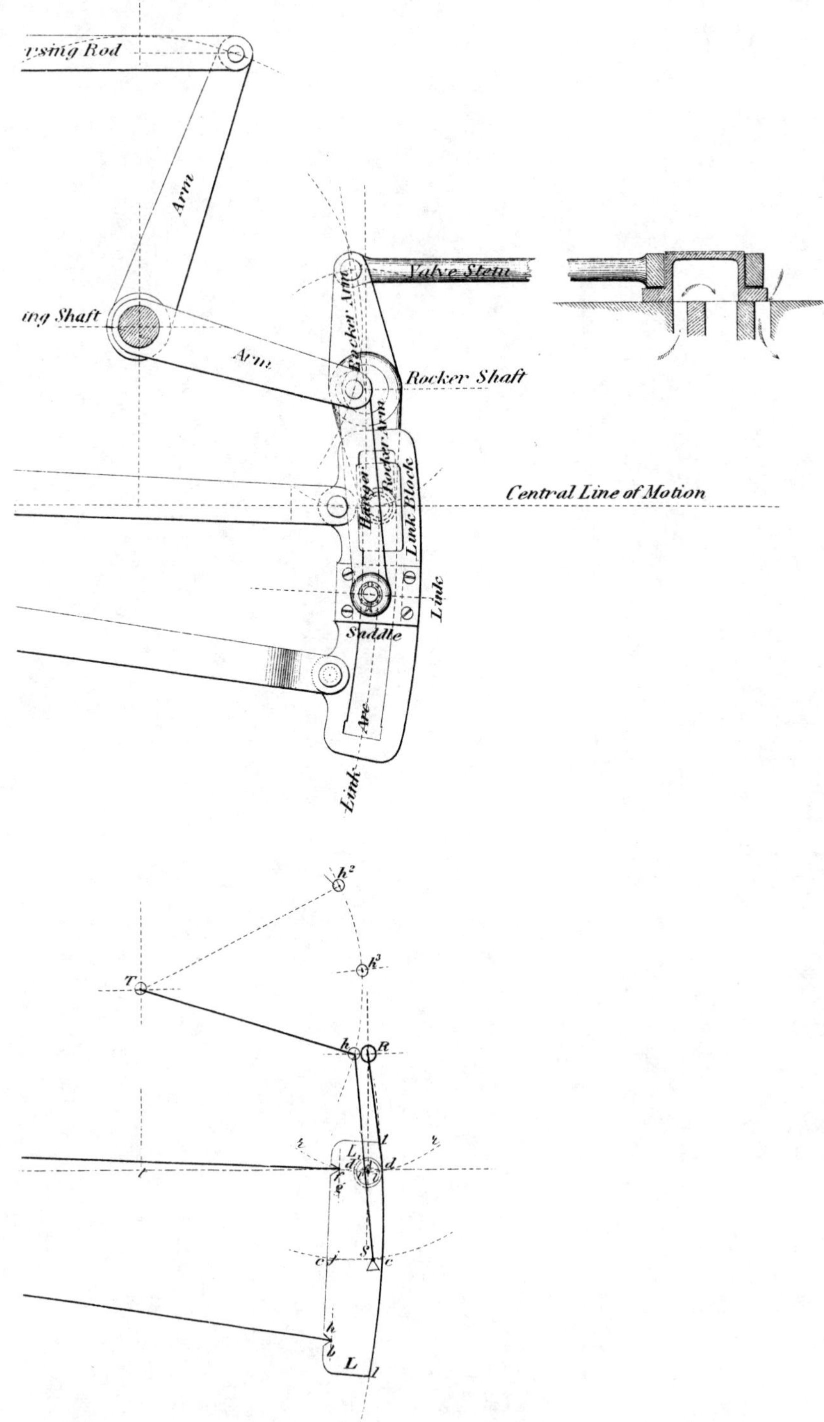
ysing Rod
Arm
ing Shaft
Arm
Rocker Arm
Valve Stem
Rocker Shaft
Rocker Arm
Hanger
Link Block
Central Line of Motion
Link
Saddle
Arc
Link
h²
h³
T
h
R
r
r
L
d
d
e
c
s
c
h
b
L
l

FIG. 24.

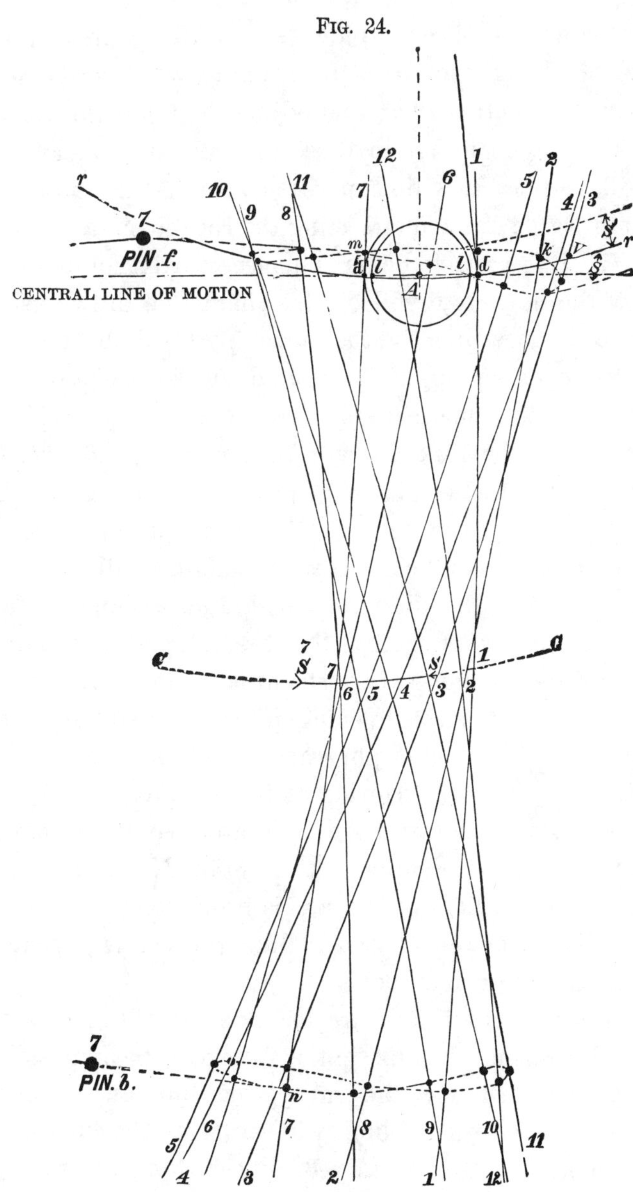

removed from their first position and the link template will
follow them with its points b and f. The only restraint

7

upon the course of this template is that the point S must travel on the hanger arc $c\,c$. If therefore we describe new arcs about the centres $f^2\,b^2$, and adjust the template so that f and b shall be found in those arcs and S in the arc $c\,c$ there will result a new link position with its arc standing like 2 2 in Fig. 24 and intersecting the rocker pin arc $r\,r$ at a point k. But as the rocker pin necessarily follows the course of the link arc it will by this change be drawn aside from d to k, consequently the steam port will be opened wider by the extent of the horizontal measurement of this distance. In like manner when the crank is carried to position No. 3 the link arc will be removed to 3 3 as in Fig. 24, and the rocker pin to V, producing thereby a still wider opening of the steam port. The same process applied to the remainder of the 12 crank positions will give the other locations of the link arc (as in Fig. 24) for the full gear of the link. Now observe that the link position 3 3 produces the widest opening of the steam port, and as the crank advances to 4 and 5 this opening grows less and less, until between 5 and 6 the rocker pin reaches the point l, where the steam is finally cut off. During its further progress expansion goes on and at last when A is attained the exhaust opens and the steam escapes. At position No. 7 (the 180° location of the crank) the link arc is brought again in contact with the lead circle and a like process is repeated throughout the return stroke.

A duplicate set of link arc locations, might readily be obtained by raising the link to the full gear back position and a similar set for the mid gear, but an examination of the one just found will develop the character of the motion. Draw curved lines* tangent to the

* Hyperbolic curves with semi-transverse axes shorter than their semi-conjugates.

extreme positions of the link arc to represent the boun-
day lines of the valve travel, as B B′ B″ and D D″ in
Fig. 25. From the centre of the main shaft, with a ra-

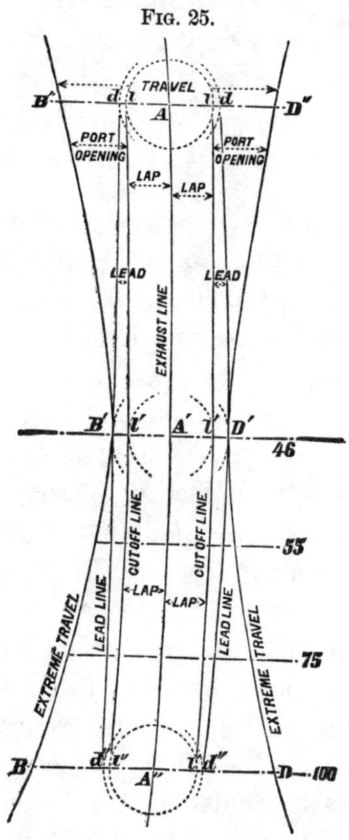

FIG. 25.

dius equal to C A sweep an indefinite arc A A′ A″ for the
exhaust-closure line, and parallel arcs l l′ l″ for the cut-off
lines.   Draw lines d D′ d″ and d B′ d″ tangent to all the link
arc positions No. 1 and No. 7; these show, by their distances
from the cut-off lines, the variable character of the lead
opening at the extremities of the stroke.

Comparing Figures 25 and 19 we find that the shifting

link, with its two eccentrics, gives results precisely analogous to those of the adjustable eccentric, consequently all the inferences (page 81) we were able to draw from the latter motion, will hold with equal force in the case of the former.

ADJUSTMENT OF LINK MOTIONS.

Besides the qualities possessed in common by the two motions, the link has that of adjustability, a very important feature, and one which specially characterizes it. As the tendency of the connecting rod angularity in a direct acting engine is to produce a *later* cut-off on the forward stroke than the amount required, and since with the link the cut-off in either stroke depends on its degree of elevation or depression; it follows that if we suspend the link in such a manner as to cause a suitable elevation for the forward stroke, the result will be a perfectly equalized motion for the gear in question. And again if the equalization be made applicable to all gears, then the link may be suspended at *any* point between the full forward and full back *without* an appreciable inequality appearing between the cut-offs or the exhaust closures of either stroke.

But a practical difficulty here arises; the link block moves upon a fixed arc $r\ r$ while the link rises and falls, consequently for each revolution of the crank *the link will slip* back and forth a certain distance on its block. Should this slip be excessive in any particular gear and the engine run a long time in this gear, the faces of the link would become worn, "lost motion" would ensue and the delicate action of the parts would be destroyed.

Hence in planning a serviceable link motion it is necessary to reduce the slip of the link to its smallest value, con-sistent with the equalization of the motion, and in *marine engines* to even sacrifice the equality of the cut-offs to the *reduction of the slip.* In Fig. 24 the motion of the two fixed points (*m* and *n*) on the link have been traced in looped curves. The upper of these, shows to what extent the point *m* falls below and rises above the arc *r r*, giving a slip equal to the distance S plus S'.

It is important to observe that the magnitude of the slip grows smaller and smaller as the link block draws nearer to the point of suspension, because this fact indicates that the stud of the saddle should be placed—when a minimum value of the slip is required at a certain *point* of suspension—as *nearly over* such point as possible.

CONNECTION OF ECCENTRIC RODS.

The variable character of the lead opening in a shifting link motion depends upon the manner in which its eccentric rods are attached, and its magnitude depends on the length of those rods. The force of this remark will appear from an examination of Figures 26 and 27. In both instances, *the eccentric centres lie between the centre of the shaft and the link*, while the latter for sake of simplicity has been made to act directly on the valve. The No. 1 position represents the mid gear, and No. 2 the full gear forward of the link. If under these conditions the eccentric rods be *crossed* as in Fig. 26 the lead opening will *decrease* from the full to the mid gear of the link, where the motion may even be without lead.

But with the *open* rods of Fig. 27 the lead opening

FIG. 26.

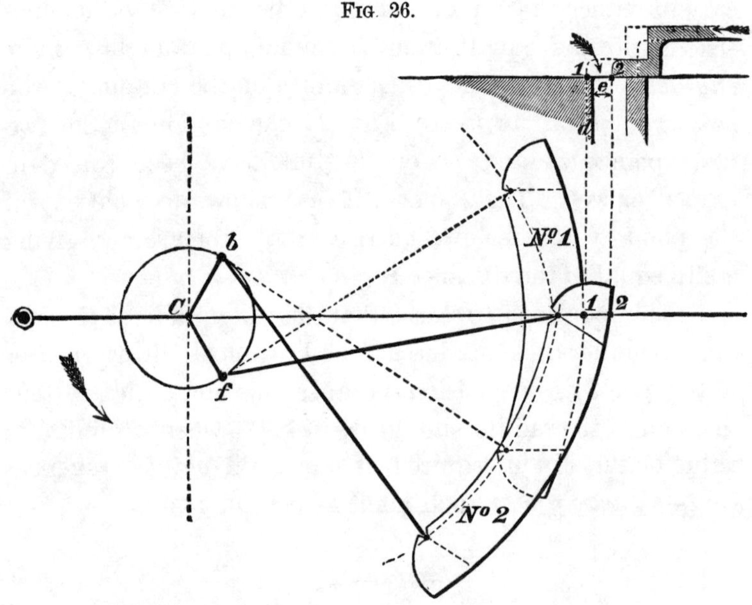

FIG. 27.

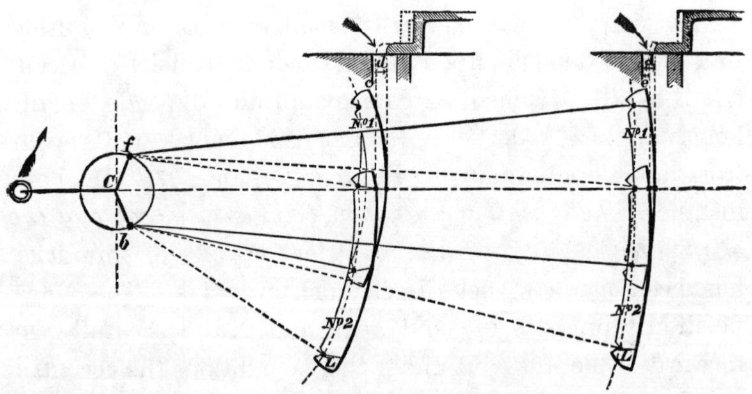

increases from the full to the mid gear, and the rapidity of
this increase, for a given link, depends directly upon the
length of the rods ; hence with a given mid gear lead open-

ing that for the full gear will be determined mainly by this length.   Excepting the case of valves having an independent cut-off (Part V.) the rods are seldom crossed as in Fig. 26, yet (from conclusion VI, page 81) it is believed that many instances exist in which the arrangement might be adopted with good results.   It is also possible, with such a motion, to stop the engine by placing the link in the mid gear ; but this can never be done with a motion like Fig. 27, whose valve is invariably opened a certain amount in the mid gear.   The extremes of mid gear lead opening in loco-motive practice are $\frac{1}{4}$ and $\frac{1}{2}$ an inch, but the more common value is $\frac{3}{8}$ inch ; while the full gear lead varies between $\frac{1}{16}$ and $\frac{3}{16}$ inch, governed principally by the length of the ec-centric rods.

With the stationary link the lead opening remains un-altered by changes in gear ; so that if $\frac{3}{8}$ inch be assumed as the proper amount for the full gears, the motion will retain this lead for all gears between these extremes and the mid gear.   This peculiarity is not inherent with the stationary link, since many *shifting link motions* may be arranged with a *Constant Lead* for the various gears of *one* direction of the motion.   Take, for example, the motion shown in Fig. 22, in which the angular advance of each eccentric equals 21° and the lead enlarges from $\frac{1}{8}''$ in the full, to $\frac{5}{16}''$ in the mid gear.   By imparting an angular advance of 31° to the eccentric F, while that of B remains unaltered, the lead opening becomes *constant* for all points between the full gear forward and the mid gear, and diminishes from $\frac{1}{16}$ inch in the mid gear to $\frac{1}{8}''$ in the full gear back.   *Vice versa* for a change in the angular advance of the eccentric B.

PRACTICAL OBSERVATIONS.

(BASED ON FIG. 22.)

I. The tumbling shaft must be located at such a distance above or below the central line of motion, that neither eccentric rod can *strike against* it when the link is moved from one full gear to the other. Special cases may arise that demand a curvature of the eccentric rod, but the practice in general should be discountenanced.

II. The hanger must be of such a length that the extremity of the link will not conflict with the tumbling shaft arm in either forward or back gear. The length of the tumbling shaft arm is usually equal to or greater than that of the hanger.

III. If the link cannot be placed in full gear back, owing to the arrest of its tumbling shaft arm by the boiler or other opposing object, either the tumbling shaft must be removed and located *below* the link motion, or the rocker must be lengthened in order to depress the central line of motion and with it, the entire motion. When the latter expedient is resorted to, a change should be made in the relative positions of the rocker arms, for the purpose of preserving the identity of their motions. The proper inclination W of the arms is found by describing a circle $r\,t\,t\,r$ (Fig. 28) tangent to the central line of the valve stem, or a line sufficiently above the same to equalize the vibration of the stem, and the central line of the motion. Radial lines from the points of tangency will then give the relative positions of the arms.

This method of correction is preferable to the former in respect to the symmetry of the motion, because the greater the length of the rocker arm, the less will be the vibration of the valve stem, as well as the slip of the link block.

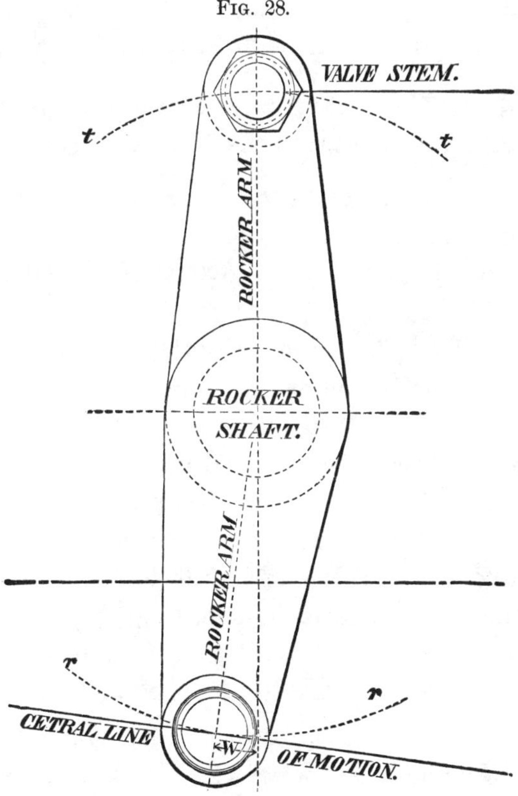

IV. So long as the angular advance of the eccentrics is laid off from a line at right angles to the central line of the link motion, the latter can be arranged at any inclination to the piston motion, without affecting the action of the link. These central lines were made to coincide in Fig. 22, merely for the purpose of simplifying the investigation, whereas they might have formed with each other any angle whatever (see Fig. 52, Part V).

GENERAL DIMENSIONS.

In ordinary locomotive practice the dimensions of the various parts range between the following extremes:

Ratio of crank arm to connecting rod $1 : 5\frac{1}{2}$ and $1 : 8$.

Travel of valve 4 to 6 inches.

Maximum cut-off from $\frac{3}{4}$ to 0.92 of the stroke (generally $= \frac{7}{8}$).

Mid-gear lead $\frac{1}{4}''$ to $\frac{3}{8}''$, usually the latter.

Full gear (dependent on length of rods) $\frac{1}{10}''$ to $\frac{3}{16}''$.

Radius of link $3'\ 6''$ to 6 ft.

Distance between eccentric rod pins $10''$ to $14''$.

Pins back of link arc $2\frac{1}{4}$ to 3 inches.

Saddle-stud back of arc $0''$ to $1\frac{1}{2}$ inches.

Stud above central line of link $0''$ to $2\frac{1}{2}''$.

Length of hanger 12 to 20 inches.

Length of tumbling-shaft arm 14 to 22 inches.

Length of rocker arm 8 to 11 inches.

The following special dimensions, collated by the Master Mechanics' Associations, are indicative of the prevailing practice on thirty-five of the railroads of our country:

Number of Roads and Class of Locomotives.	Travel of Valve..	Outside Lap......	Lead in full Gear.	Inside Lap........
	in.	in.	in.	in.
25 Roads use on their Express Pass. locomotives	5	$\frac{7}{8}$	1–10	$\frac{1}{8}$
6 " " " " "	$4\frac{3}{4}$	$\frac{3}{4}$	$\frac{1}{8}$	1–16
4 " " " " "	5	$1\frac{1}{4}$	$\frac{1}{8}$	$\frac{1}{4}$
20 " " Accomd. " "	5	$\frac{3}{4}$	1–10	$\frac{3}{8}$
10 " " " " "	$5\frac{1}{2}$	$\frac{7}{8}$	1–16	1–16
5 " " " " "	$4\frac{1}{2}$	$\frac{5}{8}$	$\frac{1}{8}$	3–16
19 " " Freight locomotives	5	$\frac{3}{4}$	1–10	1–16
11 " " " "	$4\frac{1}{2}$	$\frac{5}{8}$	1–16	$\frac{1}{8}$
5 " " " "	$4\frac{3}{4}$	$\frac{1}{2}$	1–10	3–16

GENERAL PRINCIPLES.

OF THE GEOMETRIC SOLUTION.

A cursory examination of the link motion might naturally lead to the conclusion—from the simplicity of the parts and the strong resemblance existing between their action and the single eccentric's—that the theory of the latter being perfectly comprehended, but little difficulty would attend the work of assigning proper proportions to the former. Such an inference, however, would not be strengthened by a closer inspection, much less sustained by an intelligent effort to accomplish a solution. The reason for this fact lies not only in the multiplicity of the parts, but also in the conflicting character of the elements that constitute a perfectly *equalized* link motion.

The requirements of such a motion are—*perfect equality* of cut-off, of exhaust closure, lead opening and maximum port opening, together with absence of block slip, between the forward and return stroke of the piston for *every* suspension of the link from full gear forward to full gear back. Such theoretical excellence is *absolutely* impossible with the ordinary type of link motion, and efforts made to attain the same must necessarily result in failure.

But good practical qualities *may be* obtained by *sacrificing* the *non-essential* to the *essential* points of the motion. The action of the connecting rod on a link motion, may justly be compared to the distorting effect of pressure exerted upon one point of a symmetrical india-rubber ball, producing thereby a temporary concavity. This it is true can be removed by an *even* application of additional pressure to the adjoining parts, but the ultimate effect will be a bulging out of the central portion, and the symmetry can alone be restored by withdrawing *all* pressure. Just so with the link motion, the angularity of the rod tends to render one or more events of the motion unequal in the opposite strokes of the piston, and should it appear more desirable to preserve certain ones of these than others, we must *purchase* their equality at the *expense* of the latter. Reducing the angularity of course diminishes its disturbing effect, hence in departments like locomotive engineering, where much attention is bestowed on the equalization of the motion, crank and connecting rod ratios of 1 : 7 or 8 obtain ; while in marine engineering ratios of 1 : 4 or 5 are common.

The subject of preserving the equalities of cut-off and exhaust closure at the expense of lead and port openings has been treated of in Part II. It will only be necessary to examine it here with reference to the mid gear. At this point the port and lead openings attain their minimum value, which being much less than the 0.6 or 0.9 port opening required for perfect admission, tends to reduce the pressure of the steam by wire-drawing, and if these openings vary, unequal powers will be applied in opposite strokes. Consequently the *mid gear, lead and port openings* must have *equal values* in both strokes, however irregular they may be in the full gears. No fixed limit can be assigned to the slip of the link on its block, but the amount allowable under different conditions will readily be determined by the judgment of the Engineer. In every case, the main object is to reduce the slip to a minimum value for that gear in which the engine will be most frequently operated.

GEOMETRIC SOLUTION.

LINK No. I.

In designing an engine, as a general thing, no particular part can be isolated, its proportions assigned, and its details worked out regardless of the conditions inevitably imposed upon it by the character of the adjoining parts; but rather, trial dimensions must be affixed, their adaptability tested and modified by circumstances, and finally all must be unfolded and developed in perfect harmony. When the subject of scheming the link motion comes in order, we find that peculiarities of detail have already 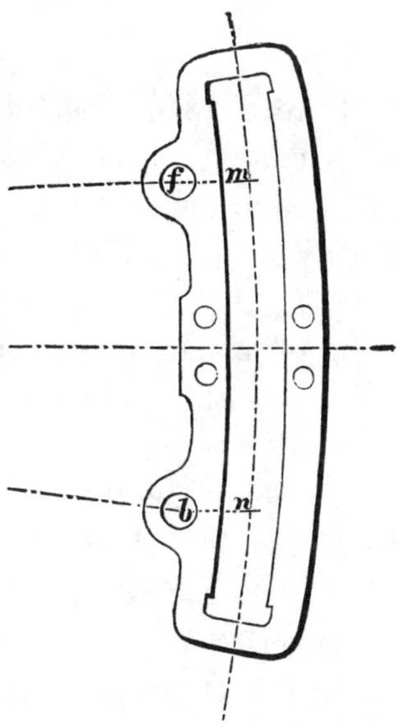 fixed the ratio of the crank arm to the connecting rod, have pointed out a convenient location for the rocker shaft and have more or less circumscribed the boundaries of the entire motion.

Since methods of construction are always most intelligibly presented when the mind is able to follow their operation in the solution of a practical example, we will take for illustration of our method the following dimensions :

Ratio of crank to connecting rod$=1 : 7\frac{1}{2}$.
Eccentric circle diameter$=5\frac{1}{2}$ inches.
Maximum cut-off$=0.92$ stroke.
Rocker from shaft$=49\frac{1}{4}$ inches.
$c.$ to $c.$ of eccentric pins$=13$ inches.
Pins back of link arc$=3$ inches.
Mid-gear lead$=\frac{3}{8}$ inch.

To find lap, full-gear lead, point of suspension of link and location of tumbling shaft.

Spread upon a long drawing board, or table, two sheets of paper large enough to contain figures similar to 29 and 31 when drawn on the Full Scale. The one will be used for locating the various important positions of the eccentric centres ; the other, for the journeyings of the link and its point of suspension, their centres should therefore be separated by the proposed distance between the shaft and rocker.

Stretch a fine thread tightly across both papers in order to locate the right line E C D A, which constitutes the *Central line of Motion.**

Describe about the point C as a centre the eccentric circle E F D, with a radius equal to the throw of the eccentric. Then, with the points of intersection E and D as centres describe with an assumed radius equal arcs intersecting at G and H and erect the perpendicular G C H to represent the neutral positions of the eccentrics. From this

----

* The use of the T square should be avoided in all of the constructions.

line lay off an angular advance appropriate to the desired cut-off. This may be found in the subjoined Table:

Cut-Off.	Angular Advance.	Return Stroke Maximum Cut-Off Angle.
0.75 = $\frac{3}{4}$	28 degrees.	124 degrees.
0.8	25 "	130 "
0.84	22 "	136 "
0.875 = $\frac{7}{8}$	20 "	140 "
0.9	17 "	146 "
0.92	16 "	148 "

In the present case the advance equals 16°, which laid off, by means of a protractor, from the line C G, determines the position F of the forward motion eccentric when the crank stands at the zero. In like manner B might be found, but it will always prove more convenient and accurate to take in a pair of dividers the distance between F and the point of intersection of the circle with the line G C and then prick off from the line G H the points b, f, B. We thus obtain the two positions F and B of the forward and backing eccentrics when the crank stands at the zero, as well as their new ones f and b for the 180° location of the crank.

It is well known that the inequality of the crank angles attains its maximum value at the $\frac{1}{2}$ stroke of the piston, hence the importance of examining the link motion with special reference to the $\frac{1}{2}$ stroke cut-off. Although appropriate angles for the crank have been furnished in the Stroke Table, it is thought best for facility of reference to here reproduce them in a more compact form.

Ratio of Crank to Connecting Rod.	Crank Angles for Half-Stroke Position of the Piston.	
	Forward Stroke.	Return Stroke.
1 : 4	$82\frac{7}{8}$ degrees.	$97\frac{1}{8}$ degrees.
1 : $4\frac{1}{2}$	$83\frac{3}{4}$ "	$96\frac{1}{4}$ "
1 : 5	$84\frac{3}{8}$ "	$95\frac{5}{8}$ "
1 : $5\frac{1}{2}$	$84\frac{7}{8}$ "	$95\frac{1}{8}$ "
1 : 6	$85\frac{3}{8}$ "	$94\frac{5}{8}$ "
1 : $6\frac{1}{2}$	$85\frac{5}{8}$ "	$94\frac{3}{8}$ "
1 : 7	$85\frac{7}{8}$ "	$94\frac{1}{8}$ "
1 : $7\frac{1}{2}$	$86\frac{1}{4}$ "	$93\frac{3}{4}$ "
1 : 8	$86\frac{3}{8}$ "	$93\frac{5}{8}$ "

If the ratio of crank to connecting rod be that of $1 : 7\frac{1}{2}$ the two eccentrics will advance $86\frac{1}{4}°$ from F and B while the piston travels to its forward $\frac{1}{2}$ stroke location, and $93\frac{3}{4}°$ from f and b for the return stroke.

A very convenient way of locating these four points is to lay off from C F by means of a protractor, the point $\frac{1}{2}f$ distant $86\frac{1}{4}°$, and $f\frac{1}{2}$ distant $93\frac{3}{4}°$ from fC. Then with a pair of dividers prick off these points at equal distances on the opposite side of E and D giving thereby the two other points $\frac{1}{2}b$ and $b\frac{1}{2}$. In the nomenclature here adopted the letter f refers to that eccentric which produces a forward or positive motion of the crank, while b always designates the eccentric for the back or reverse motion.

When a fraction is prefixed to either of these letters, it signifies some forward stroke position of the piston with the link in the forward gear, but if it follows the letter a return stroke of the piston with the link in the same gear. Thus, the $\frac{1}{2}f$ represents the position of the forward eccentric when the link is in the forward gear and the piston has advanced to its forward $\frac{1}{2}$ stroke location, and $f\frac{1}{2}$ represents the same when the piston has attained its $\frac{1}{2}$ stroke return position. Having accurately located these eight important positions

[RATIO, 1:7½.]
G.
16°
b.
F.
Eccentric Rod.
86¼.
b¾.
E.
C.
D.
Central Line of Motion.
3f.
4b.
93¾.
f.
B.
Eccentric Rod.
H.

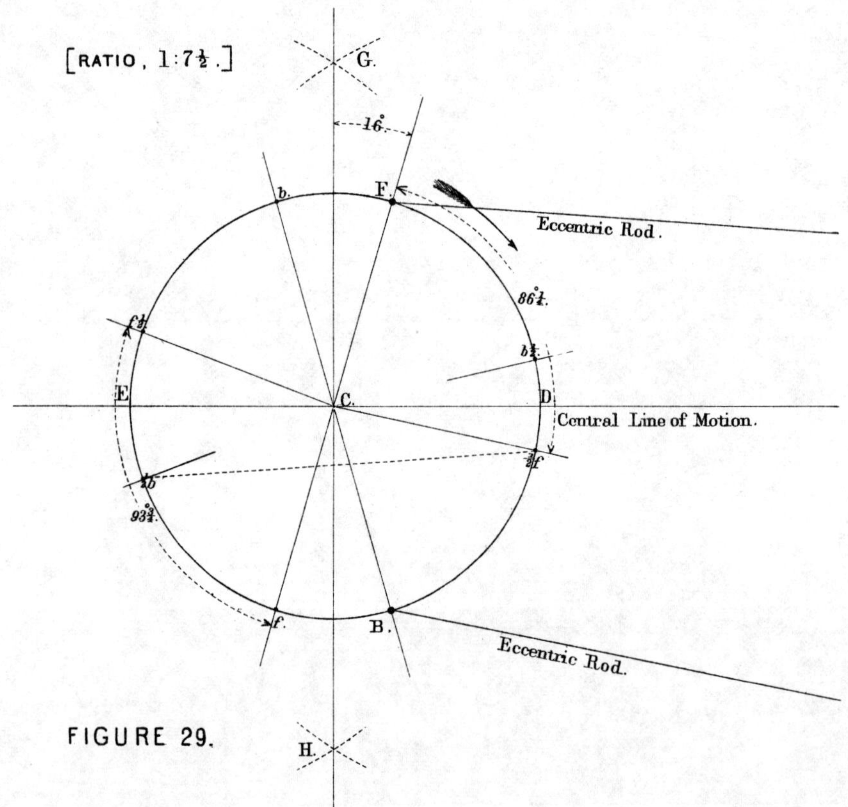

FIGURE 29.

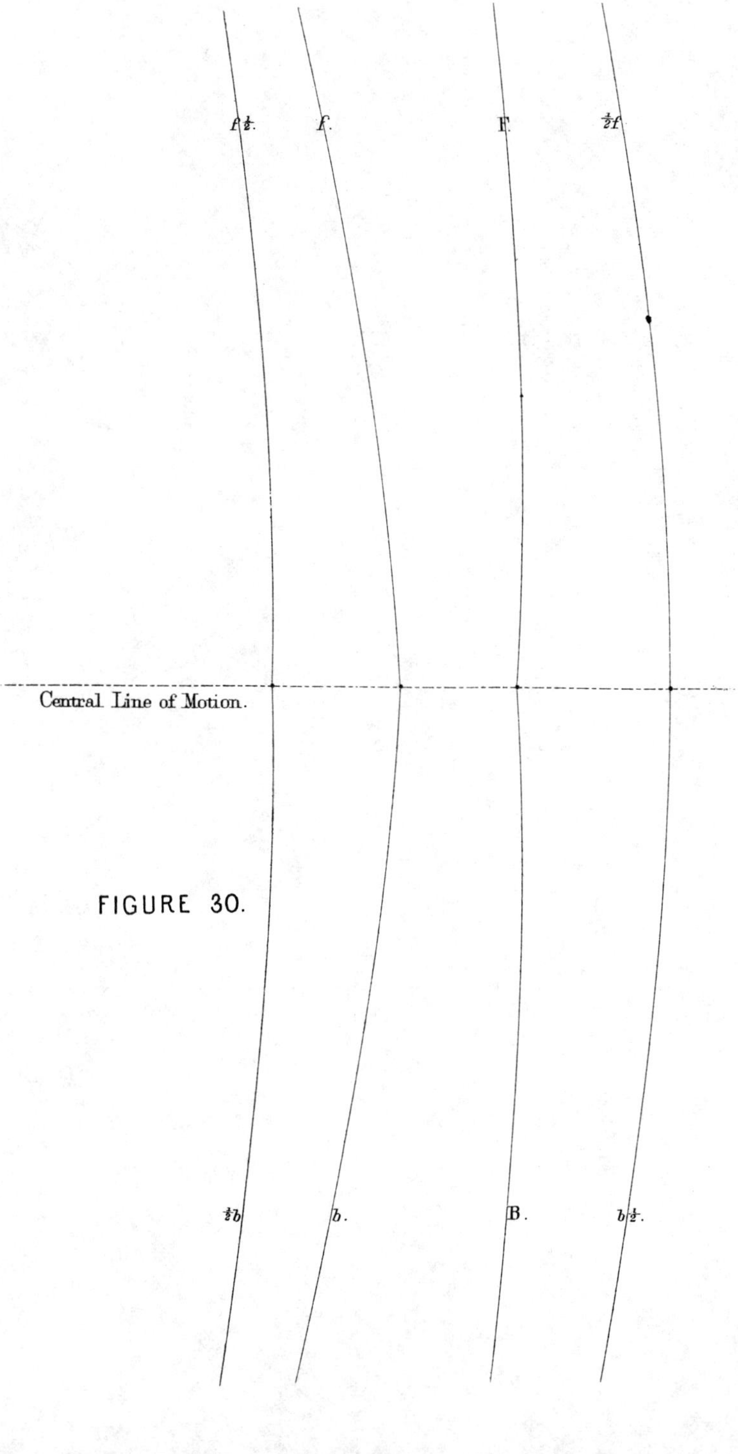

FIGURE 30.

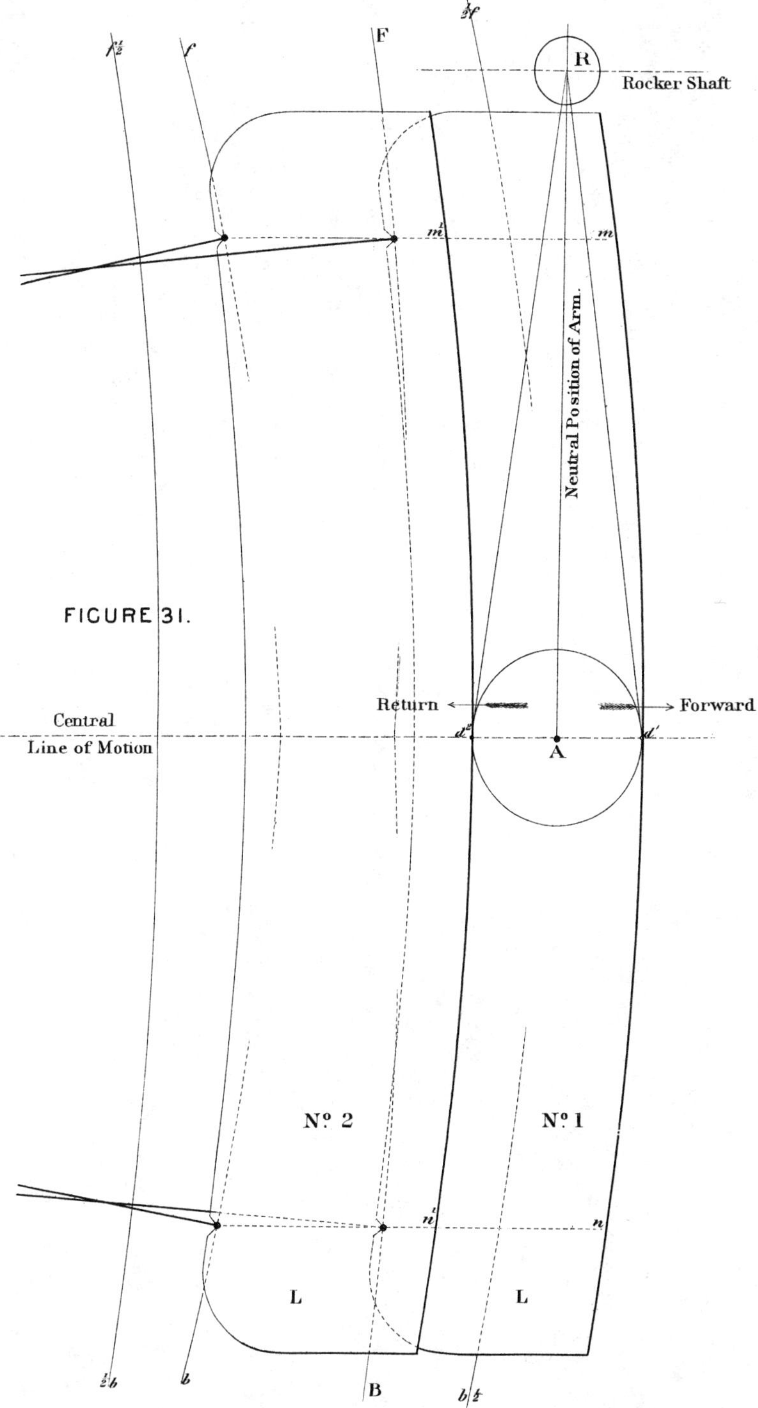

Rocker Shaft
R
Neutral Position of Arm.
Central Line of Motion
Return
Forward
A
FIGURE 31.
No 2
No 1
L
L
F
B
m'
m
n'
n

of the eccentrics, we pass to the other sheet of paper and trace their influence on the proportions of the link attachments.

Since the assumed distance of the rocker from the shaft is $49\frac{1}{4}''$ and the eccentric rod pins are withdrawn $3''$ back of the link arc, the length of each rod will equal $49\frac{1}{4}'' - 3'' = 46\frac{1}{4}''$. Adjust a pair of beam compasses to strike arcs of $46\frac{1}{4}''$ radius. Step the needle point successively in the eight locations of the eccentric's centres just found, and sweep from the central line of motion the same number of indefinite arcs (as shown in Fig. 30) upon which the eccentric rod pins must *inevitably* travel for the four given positions of the crank arm.

Next, from a piece of white holly veneer cut a template L (Fig. 32) having a link arc of $49\frac{1}{4}''$ and with $\vee$ incisions, $13''$ apart and $3''$ back of the arc to represent the location of the eccentric pins, and draw upon the same the three parallel lines $f\,m, j\,$S$, b\,n$; making $j\,$S lie midway between f and b as well as perpendicular to a line joining these points. We are now prepared to trace the journeyings of the link arc.

I. To find the mid-gear travel.

For this purpose place the template in the mid gear positions No. 1 and 2 (Fig. 31) with its eccentric pins on the arcs F, B, f, b, and mark the points d', d^2 in which the link arcs intersect the central line of motion. Locate these points permanently by describing a circle through them, having its centre A in the central line. This point gives us the *true* location for the rocker shaft, the distance from the main shaft being equal to C A instead of $49\frac{1}{4}''$.* Through A, therefore, erect a perpendicular A R to the central line of motion and on it locate the centre R of the rocker shaft.

---

* Their difference is always so trifling that the rocker box may readily be moved the proper amount and much difficulty of construction be thereby avoided.

II. To find the lap of the valve.

From d^1 and d^2 lay off the mid-gear lead opening of $\frac{3}{8}$ inch towards A, and permanently locate the positions thus found by sweeping about the centre A a second circle $l\,l'$. But since the mid-gear travel invariably equals the sum of the laps plus the mid-gear lead openings, the diameter of the circle $l\,l'$ will equal the sum of the laps, consequently the simple lap of the valve must equal its radius $A\,l$ or $A\,l'$, Fig. 32.

The following Table will aid the designer in the selection of a suitable lead opening after the mid-gear travel has been determined. For since the value of the lead angle may range between about 30° and 40°, the widths of the openings will be those found in the Table. Of course much latitude is here allowed to the exercise of individual judgment, for the subject demands it. Observe, the larger travels are only found on marine engines:

MID-GEAR DIMENSIONS.

Mid-gear Travel.	LEAD OPENING FOR A LEAD ANGLE OF—		
	30°.	*35°.*	*40°.*
1 inch.	$\frac{1}{8}$ inch.	$\frac{3}{16}$ inch.	$\frac{1}{4}$ inch.
$1\frac{1}{2}$ "	$\frac{3}{16}$ "	$\frac{1}{4}$ "	$\frac{5}{16}$ "
2 "	$\frac{1}{4}$ "	$\frac{3}{8}$ "	$\frac{7}{16}$ "
$2\frac{1}{2}$ "	$\frac{5}{16}$ "	$\frac{7}{16}$ "	$\frac{9}{16}$ "
3 "	$\frac{3}{8}$ "	$\frac{1}{2}$ "	$\frac{11}{16}$ "
$3\frac{1}{2}$ "	$\frac{7}{16}$ "	$\frac{5}{8}$ "	$\frac{13}{16}$ "
4 "	$\frac{1}{2}$ "	$\frac{11}{16}$ "	$\frac{15}{16}$ "

III. To find position of the stud for equal cut-offs at the $\frac{1}{2}$ stroke of the piston.

Place the template in position No. 3, with its eccentric pins on the forward $\frac{1}{2}$-stroke elements $\frac{1}{2}f\,\frac{1}{2}b$, and its link arc in contact with the lap or cut-off point l. Then mark upon the paper the position j S occupied by its central line

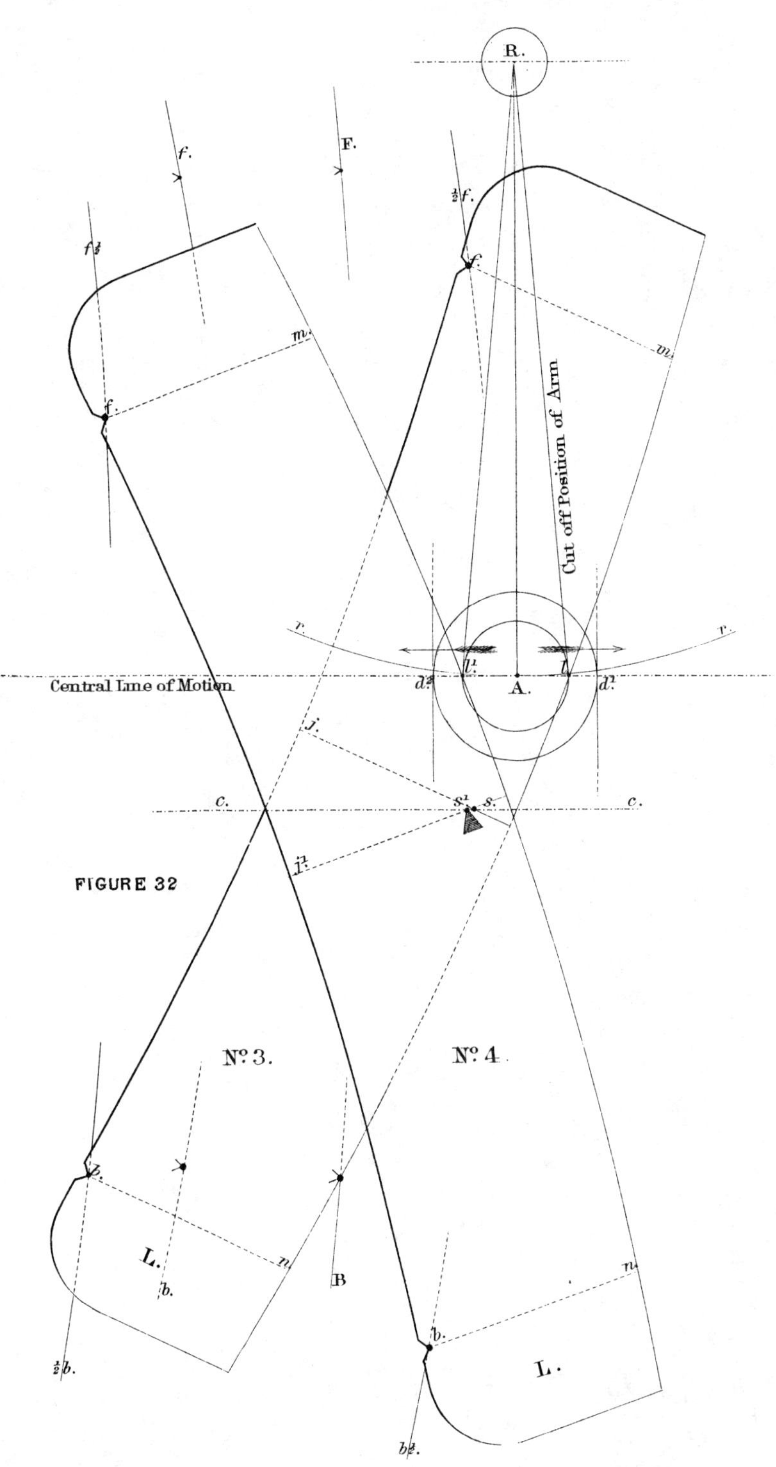
R.
f.
F.
½ f.
f ½
Cut off Position of Arm
m.
m.
f.
f.
r.
r.
l'.
l.
Central Line of Motion
d'.
A.
d'.
j.
s'. s.
c.
c.
i'.
FIGURE 32
N° 3.
N° 4.
b.
L.
n.
n.
b.
B
½ b.
b.
b ½.
L.

together with a portion of the link arc. Next, place the template in the position No. 4, with its pins on the arcs f $\frac{1}{2}$, b $\frac{1}{2}$ and link arc over the other cut-off point l', after which mark on the paper the second position, j' S', of the link centre line. Having decided to suspend the link centrally, the point of suspension must be found on the line j S, and considering the manner in which it hangs from the tumbling shaft it is evident that for a short distance the stud will practically move along some straight line c c *parallel* to the central line of motion. The point of suspension therefore must reside in the central lines j S, j' S', must be equally remote from the link arcs and at such a distance that a line drawn through the two points will prove parallel to the central line of motion. The only two positions satisfying such conditions are S and S', found by trial distances laid off with a pair of dividers from the two link arcs. Having secured the proper distance for the stud, fix it permanently, by making a V incision in the link template; for as our subsequent study of the link will be intimately associated with the motion of this point, it is important to be able to mark its position for other gears of the link.

The inequality of the crank angles for different positions of the piston attains its maximum value at the $\frac{1}{2}$ stroke and gradually fades out at the extremities, and since we have equalized the cut-off for the $\frac{1}{2}$ stroke, it only remains to perform the same office for the maximum cut-off before we practically equalize the motion for ALL *intermediate gears* between the full and mid gears. Our next step, therefore, will be to return to Fig. 29 and map on it the four positions of the eccentrics for the maximum cut-off. The third column of the Angular Advance Table gives the maximum cut-off angle for the *return* stroke, which in the present instance=148°, and the STROKE TABLE shows that the forward stroke angle is $4\frac{1}{4}°$ less than the return, or $=143\frac{3}{4}°$. With these angles known, the four positions for the maximum cut-off may be readily laid off with a protractor, but the nature of the case enables us to present a more rapid solution. From C F (Fig. 33) lay off an angle of $143\frac{3}{4}°$. This

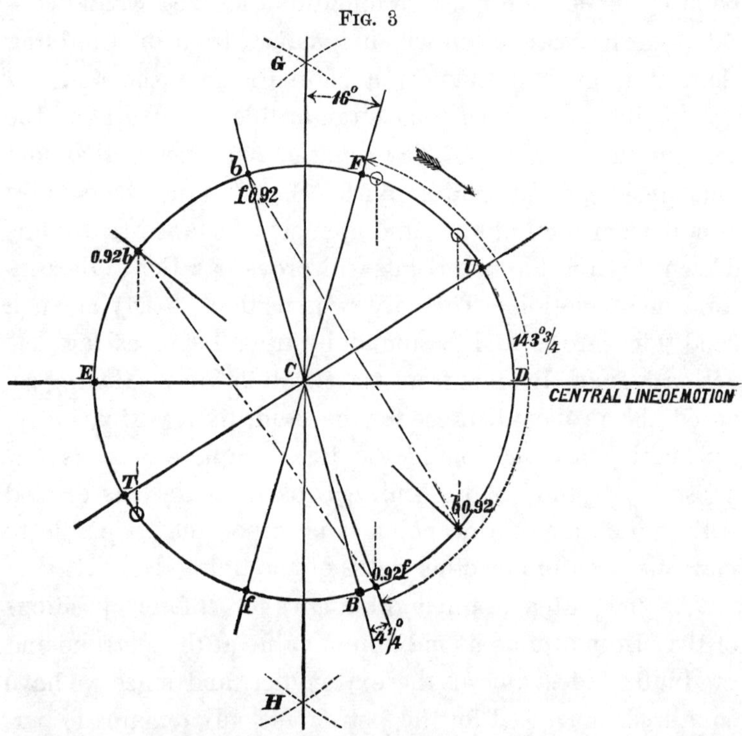

FIG. 3

locates the forward eccentric position $0.92f$ in the forward stroke. On the return stroke the same eccentric will be found at the old point b. Take the distance $0.92f$ from f, in a pair of dividers and lay it off from b in order to find $0.92b$. In like manner take that from f to B and lay it off from B, giving thereby $b\,0.92$ the last one of the four maximum cut-off points sought. With these points as centres and the length of the eccentric rod as radius sweep indefi-

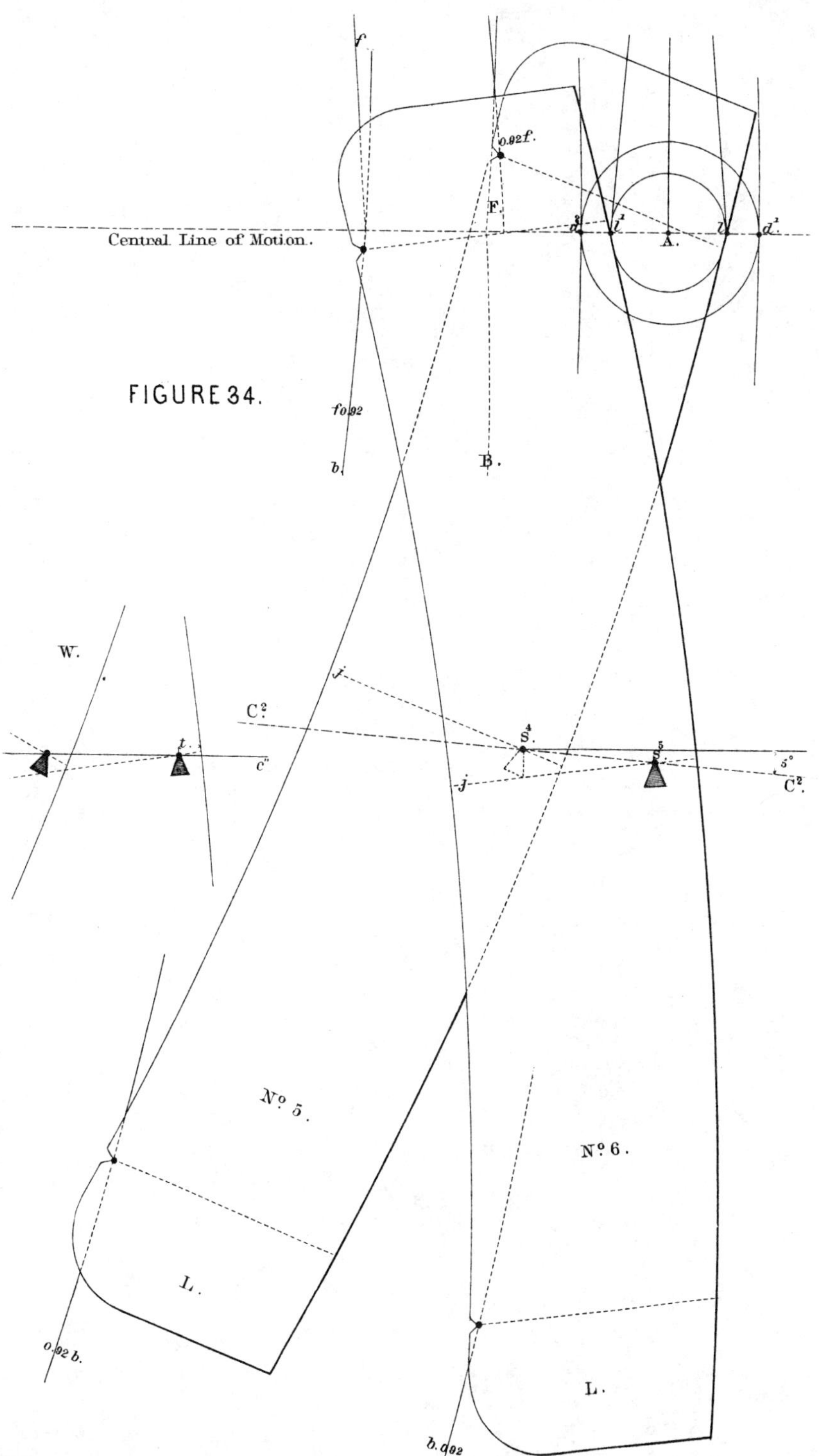

Central Line of Motion.
FIGURE 34.
o.92 f.
F
A
d
l'
l
d'
B
fo.92
b
W
j
C²
S
S
j
c''
t
C²
5°
No. 5.
No. 6.
L.
L.
o.92 b.
b.o92

h^5
h^3
h^2
h^1
h
h^4
T
Tumbling Shaft Arm.
T^2
Hanger.
C^3
s^6
s^7
C^3
C^1
s^2
s^3
C^1
d^2
l^1
l
d^2
A
Central Line of Motion.
C
s^1
s
C
C^2
s^4
s^5
C^2

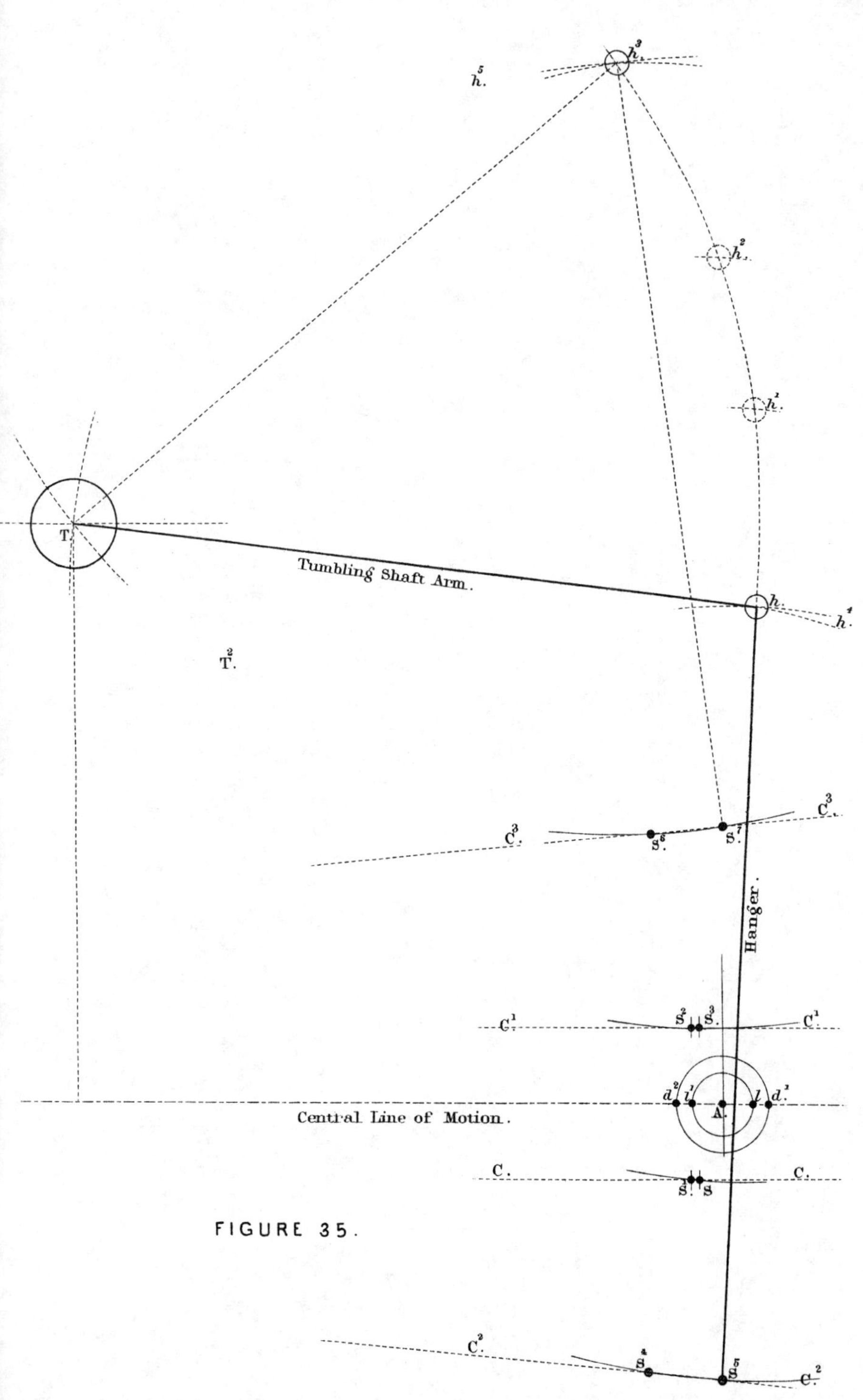

FIGURE 35.

nite arcs (as shown in Figure 34), on which the link template may travel in the full gear.

IV. To LOCATE THE TUMBLING SHAFT FOR ACCOMPLISHING AN EQUALIZED CUT-OFF IN ALL GEARS.

Slide the link template with its eccentric rod pins on the elements 0.92 f; 0.92 b, until its link arc comes in contact with the lap or cut-off point l (position No. 5) and mark the point S^4 occupied by the stud.

Again, slide the template on the return stroke elements f 0.92; b 0.92 until its link arc is in contact with the other lap point l', and mark the stud position S^5. Join the points S^4 S^5 by a line c^2 c^2, which is found to have an inclination to the central line of motion of about 5° instead of parallelism* as with c c.

By projecting the eccentric position points to the opposite side of their circle, sweeping indefinite elementary arcs with the eccentric rod as radius, and applying to them the link template, a corresponding set of stud locations S^2, S^3, S^6, S^7 (Fig. 35) may be found for equal cut-off in the back gear. But such efforts are uncalled for in the class of motions just described, because their back motion will be a precise counterpart of their forward motion, consequently the latter may be reproduced from the former, as in Figure 35.

Having thus determined 8 positions of the centre of suspension for equal $\frac{1}{2}$ strokes and maximum cut-offs, it only remains to sustain the hanger in such a manner that for the different elevations it will sweep arcs passing through ALL of these points. Arcs of intersection formed with an assumed length of hanger as radius and these points as cen-

---

* Parallelism might be secured by moving the stud S to within a distance t of the link arc (see Fig. W), but such a change would destroy the equality of the $\frac{1}{2}$ stroke cut-offs.

tres will locate the points h and h^3, and in like manner the tumbling shaft arm will determine its centre of shaft T.*

V. To find the lead of the forward and return strokes in the full gear.

Having swept with the hanger an arc $c^2\,c^2$ (Fig. 36) upon which the stud travels in the full gear of the link, slide the template on the forward lead elements F, B, until its stud lies at S^7 in the full gear arc $c^2\,c^2$, and mark the point d in which the link arc then intersects the central line of motion.

In like manner slide the template on the return stroke elements f, b, and mark the intersection d^3.

The distances of these points from the lap circle will equal their respective leads. Thus in the forward stroke the lead equals $l\,d$ in the full gear, but $l\,d'$ in the mid gear ; while $l'\,d^3$ equals the lead opening of the full gear return stroke, but $l'\,d^2$ of its mid gear. In the present case both mid-gear leads were made equal to each other. Their slight variation in the full gear has absolutely no effect on the motion.

VI. Extreme travel and slip of the link.

Referring to Fig. 34 we observe that the forward eccentric attains the extreme points of its throw at D and E on the central line of motion at which times the backing eccentric occupies the positions T and U. [The latter points may be laid off from D and E with a pair of dividers set to the distance F b.] By sweeping the elementary arcs of the eccentric rod pins for these points and adjusting the template thereto, we obtain the positions Nos. 9 and 10

---

* If the ratio of crank to connecting rod had been that of 1 : 5 or 6, the lines $c^2\,c^2$, $c^3\,c^3$ would have had a greater inclination to the central line of motion, thereby removing h and h^3 to h^4 and h^5, and depressing the shaft to some impracticable point T^2, where it would have been brought in contact with the forward eccentric rod when the link was in the back gear. The proper adjustment for such a case will shortly receive our attention.

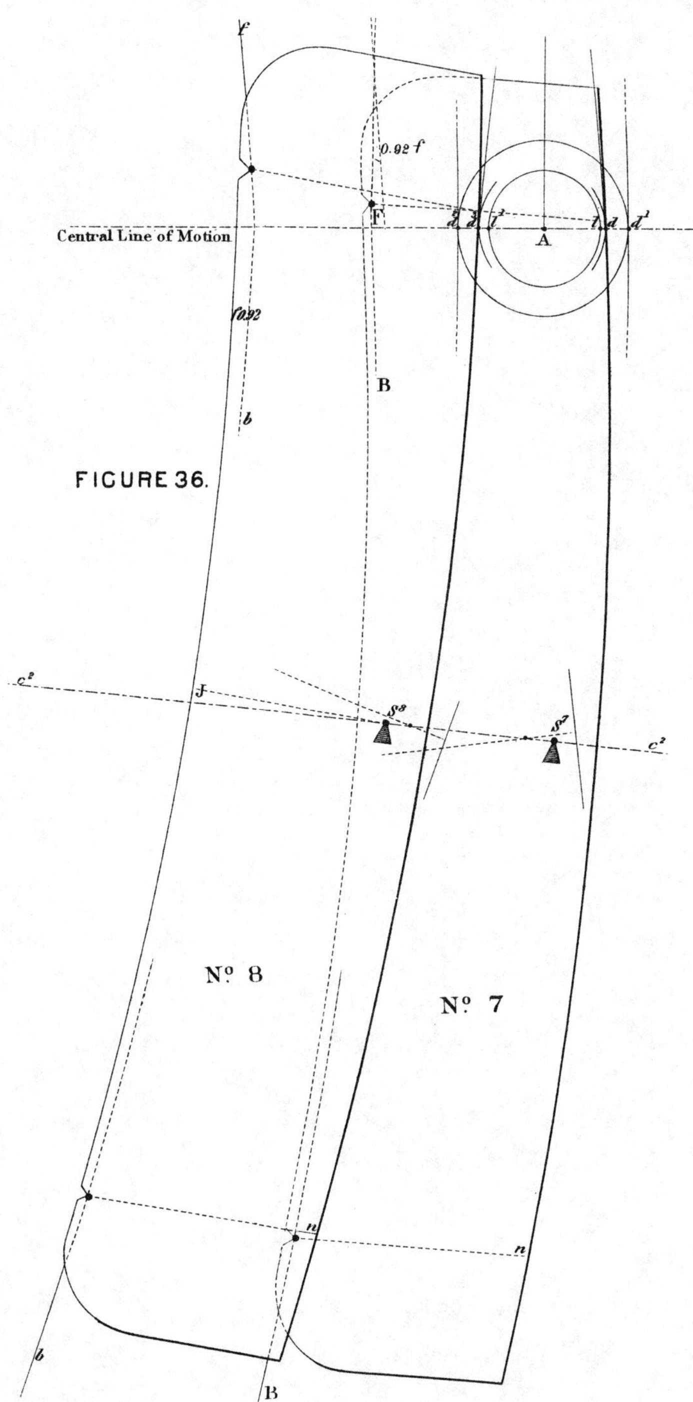

f'
0.92 f'
Central Line of Motion
F
z z t² A t d d¹
0.92
b
B
FIGURE 36.
c²
J
s8
s7
c²
N.° 8
N.° 7
n
n
b
B

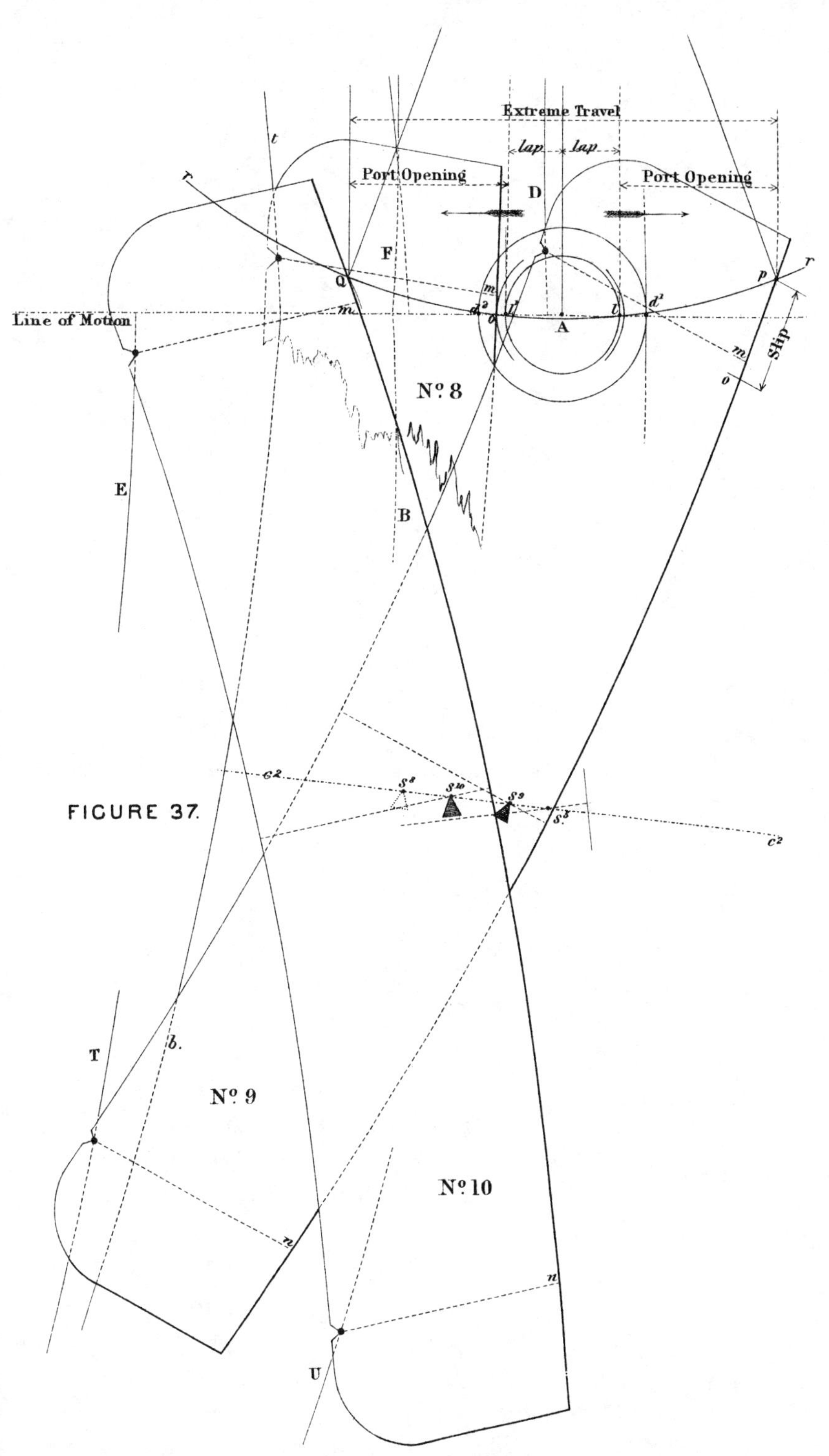

Extreme Travel
lap
lap
Port Opening
Port Opening
D
F
r
t
Q
m
m
d²
o
l³
A
l²
d³
p
r
Line of Motion
Slip
E
B
No. 8
FIGURE 37.
c²
s⁸
s¹⁰
s⁹
s⁵
c²
T
b.
n
No. 9
No. 10
n
U

(Fig. 37) and are able to mark the extreme points Q, *p*, of the rocker arc which are separated by a horizontal distance equal to the extreme travel.

For position No. 8 the fixed point *m* on the template attains its maximum elevation above the link arc, and now, at the extreme throw, its greatest depression below that arc. The maximum slip will consequently equal the distance from *p* to *o* on the link arc. This slip grows less and less the *nearer* the stud approaches the rocker pin, and if the intention should be to use the link principally in the $\frac{1}{2}$ gear this amount of slip would not prove detrimental.

MODIFICATIONS.

We have thus far, as concisely as possible, presented a geometric method for determining the proportions of all link motions, similar to those illustrated in Fig. 22. But its application cannot be considered *universal*, until certain expedients are explained, by which some of the results may be varied at will and also the motion corrected, when the ratio of crank to connecting rod is other than that of $1 : 7\frac{1}{2}$.

I. How to reduce the slip.

In the link motion of Fig. 37 the greatest slip occurs in the full gear and the least in the mid gear. Now it frequently happens that after designing a motion the maximum slip is too great for practical purposes and the query arises : what change should be made for effecting its reduction ?

There are four varieties of alterations capable of accomplishing this object, which we will here mention in the order of their relative efficiency.

1st. *Increase the Angular Advance.*
2d. *Reduce the Travel.*
3d. *Increase the Length of the Link.*
4th. *Shorten the Eccentric Rods.*

Any one or more of these agencies may be employed at the discretion of the designer and a more perfect motion be produced. Thus we could diminish the slip to $\frac{1}{2}$ of its present value, by either increasing the angular advance to 30°, or by reducing the travel to 4 inches. Of course any such change involves an entire reconstruction of the motion in accordance with the principles already explained.

II. How the slip may be distributed.

Referring to Fig. 24 we observe that the general tendency of the fixed point m is to move in an arc the *reverse* of that pertaining to the rocker pin, while n traverses one more or less parallel. The maximum slip of the forward gear consequently exceeds that of the back gear. These quantities can be equalized in great measure by placing the stud, not on the central line j S between f and b but, upon some parallel line nearer to f than to b, usually from 2 to $2\frac{1}{2}$ inches above j S. In such case the proper location for the stud is found by first drawing such a suspension line upon the link template, then sliding the template to the positions Nos. 3, 4, 5 and 6 for the forward motion, and locating the line in question at each of these positions. Make similar locations (found with the proper arcs) for the back motion. Finally locate the stud for the full gear forward in such a manner that the line c^2 c^2 joining its points shall be parallel to the central line of motion. Two or three trials may be necessary before a suitable height for the suspension line above j S is obtained. This mode of suspension is frequently adopted on locomotive link motions.

On the other hand, where no importance is attached to

the accuracy of the back motion, the slip may be greatly diminished by inclining the rocker arms to each other (Fig. 28) so that the arc $r\ r$ of the pin shall, instead of preserving a state of tangency to the central line of motion, intersect it in a similar manner to the path of the point m, Fig. 24. This method can be employed with advantage in designing marine link motions.

III. SHORT CONNECTING RODS.

If the ratio of crank to connecting rod had been assumed at $1 : 4\frac{1}{2}$ instead of $7\frac{1}{2}$, the position of the stud S found as before for equalized cut-off at the $\frac{1}{2}$ stroke, and the template slid to the positions 5 and 6, the change would have impressed on the line $c^2\ c^2$ an inclination of 18° (instead of 5°) to the central line of motion. This would have rendered it impossible to equalize the motion in the ordinary way for more than *one direction* of the crank, without bringing the tumbling shaft and eccentric rods in conflict. But if the aim be simply to equalize the forward without regard to the back motion—*a very common practice*—no special difficulty will be experienced in hanging the tumbling shaft to even this case, for the point h^3, Fig. 35, would then be left out of the account.

There exists, however, a method of equalization which corrects the difficulty for both the forward and the back gears. It is clear that the tumbling shaft is employed most conveniently and successfully, when it sustains the hanger in such a manner as to *guide its vibrations in arcs practically parallel to the central line of motion.* Hence if we wish the link to conform with this condition it will be necessary to raise the template from position No. 6 of Fig. 34 to No. 11 of Fig. 38 in which the line $c^2\ c^2$ becomes *parallel* to the central line of motion. This elevation moves the lap point l' to l^2 giving thereby a smaller lap circle $l\ l^2$ from which to

determine the lap, and at the same time increasing the lead
opening of the return stroke. The position A of the rocker
is thereby carried to a point a more remote from the shaft,
and the lead opening of the return stroke in the mid gear
becomes greater than that for the forward gear. But we
have already seen that whatever inequalities of lead opening may arise in the full gear, none can be tolerated in the
mid gear. Nor is there an occasion for their existence, because the *link arc* may be struck with a SHORTER *radius*
than the distance from the shaft to the rocker and all such
inequalities be entirely eliminated.

Take in a pair of compasses, the radius A d' of the mid-
gear travel and strike the circle d^4 d^5 about the new
position a of the rocker. To this the link arc must be
tangent when the template is placed at the mid-gear position
No. 1 and 2. Bring the template to position No. 2, mark
the fixed points m and n of the full gears on the paper, and
then search for a radius, having its centre in the central line
of motion, whose arc shall embrace the three points
m, d^5, n.

In the present extreme case the radius equals $41\frac{1}{2}''$
against $49\frac{1}{4}''$ employed with the ratio of $1 : 7\frac{1}{2}$. The two
little cuts V and Z, Fig. 39, illustrate the character of the
lead openings before and after the change of the link arc
radius.

Having thus determined the *true radius* for the link arc
a *new* template should be constructed, and all the locations
made which are appropriate to the template positions Nos.
1, 2, 3, 4, 5, 6, 7, 8, 9, 10, under the new conditions, just as
though no previous investigation had taken place.

The result will be a motion capable of ready suspension
from a tumbling shaft, with perfectly equalized cut-offs,
with port openings varying to a slight extent in the full

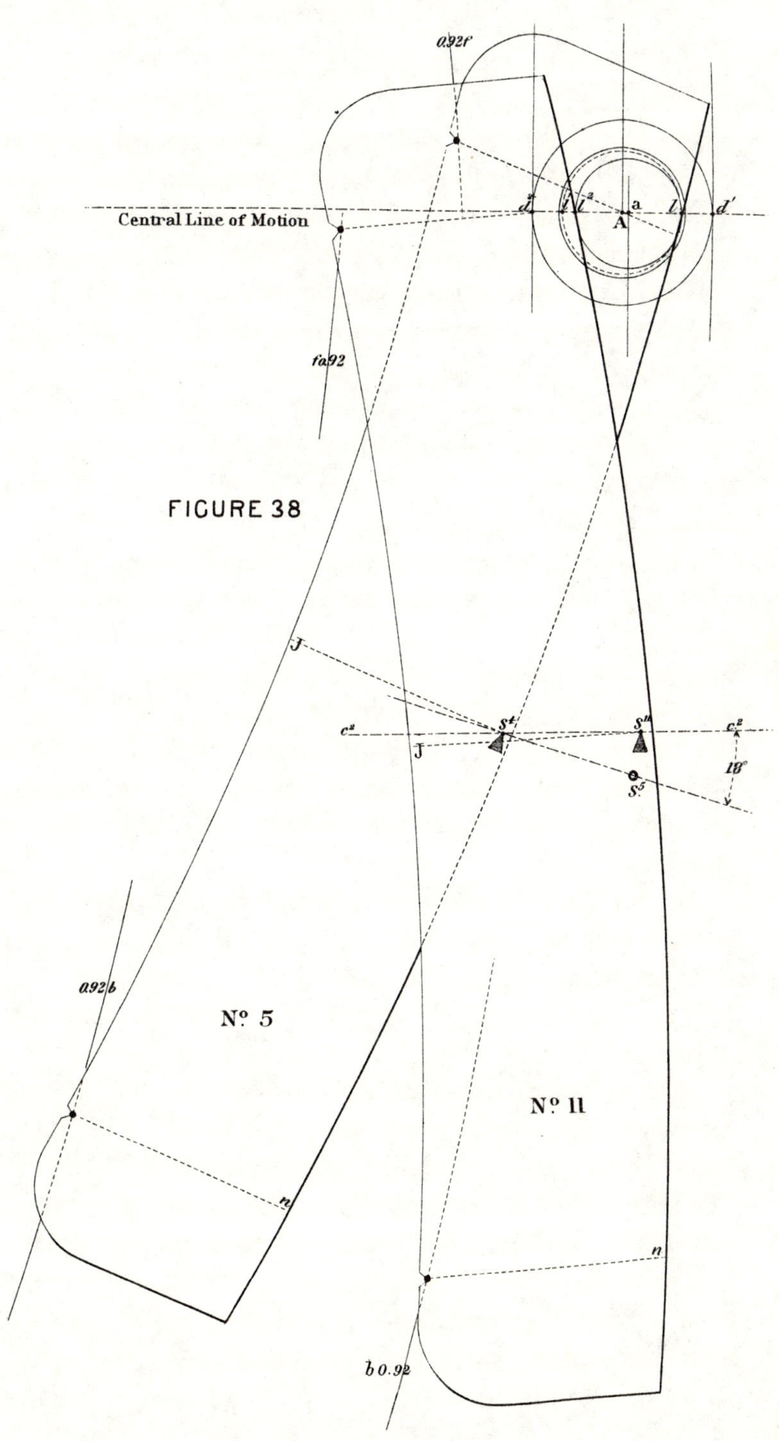

Central Line of Motion
0.92f
fa92
FIGURE 38
0.92b
No 5
b 0.92
No 11
J
J
c²
c²
S'
S''
S⁵
18°
d²
l l²
a
l
d'
A
n
n

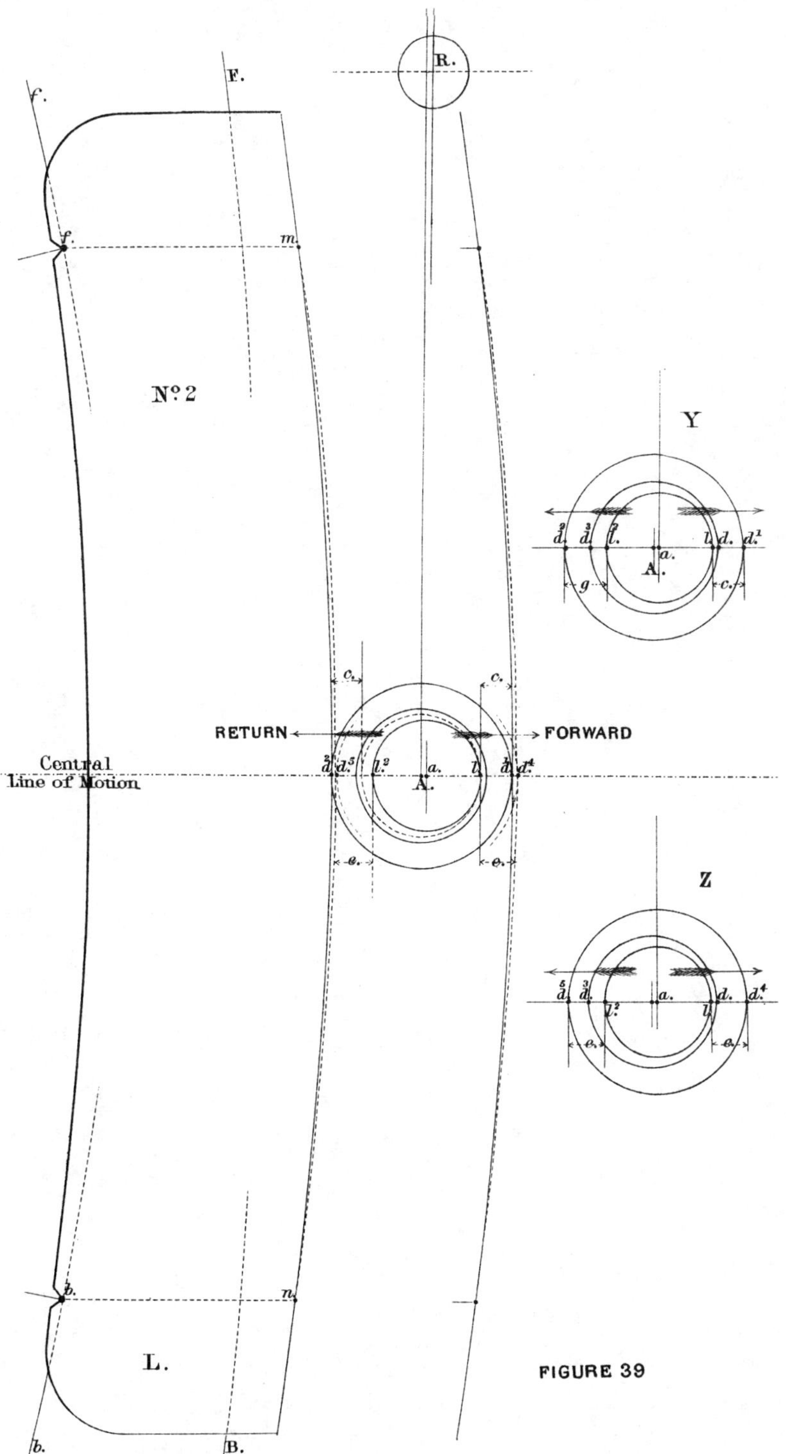

F.
R.
c.
N.º 2
f.
m.
Y
d. d. l. a. l. d. d.
A.
g.
c.
c. c.
RETURN
FORWARD
Central
Line of Motion
d. d. l. a. l. d. d.
A.
e. e.
Z
d. d. a. d. d.
l.
e. e.
b.
n.
L.
b.
B.
FIGURE 39

gears but precisely equal in the mid gear, with a forward stroke lead opening increasing from $l\,d$ (Fig. Z) in the full to $l\,d^4$ in the mid gear, and with a return stroke lead, opening gradually from $l^2\,d^3$ in the full to $l^2\,d^5$ in the mid gear. Both these lead openings will be *exactly* equal to each other in the latter gear. It will be observed that all the inequalities of the motion are thus brought in the full gears, the very position where their influence is the least injurious.

Of course it is not pretended that great inequalities of port openings are admissible in the full gears, unless the smallest of these openings shall exceed the 0.6 or 0.9 area (mentioned on page 23), in which case the magnitude of their difference becomes a matter of little importance, because the wide opening will then admit no more steam than the narrow. But since the 0.6 or 0.9 area must be reached before the mid gear is attained, where the areas become equal, it is desirable to have as little irregularity in the other openings as possible.

In reference to the lead opening, it is not unusual in locomotive "stationary-link" motions to give a *constant* lead of $\frac{3}{8}$ inch, where one increasing from $\frac{1}{16}{}''$ or an $\frac{1}{8}{}''$ in the full gear to $\frac{3}{8}{}''$ in the mid, would be employed under like circumstances for a shifting link motion, and so far as can be discovered both motions are productive of equally good results. Hence we fail to perceive the force of objections so frequently urged against a slightly irregular lead, but believe that within proper limits resort should be had to this *most efficient means* for correcting the inequalities due to a short connecting rod. The recommendation, however, must be qualified for marine engines, whose more massive reciprocating parts require *equal* admissions for bringing them smoothly to a state of rest at the extremities of the stroke. For all such it is better to equalize the lead openings, at least in the full gear, and as far as practicable the cut-offs for the same, paying but little attention to the back gear.

It has doubtless been observed that throughout the pre vious investigation no allusion has been made to the equality of the exhaust closure, and no direct effort put forth for its conservation. The omission was intentional, simply because the very process of equalizing the cut-off incidentally accomplished this object. In proof of which, it is only necessary to consider that the neutral position (A) or exhaust-closure point lies *exactly midway* between the cut-off points l, l', so that if the motion be corrected for the latter when these are near each other, it must become practically perfect for the former. But if the maximum cut-off should take place at about the $\frac{3}{4}$ stroke of the piston, the lap points l and l' would be widely separated, and probably give rise to a marked inequality of the exhaust closure at the mid gear. Since irregularities of closure and release produce a greater impression on the motion when they occur at the mid rather than at the full gear of the link, and since it is possible by means of inside lap and clearance to correct them for either of these gears, it will be well to determine the extent of the inequality for the *mid gear*, and regulate accordingly the position of the exhaust chamber with reference to the edges of the valve. To determine this correction, bring the link template to one set of the half-stroke elements (Fig. 32) and slide it thereon until the link arc stands over the exhaust-closure point A (or a, as the case may be). Mark the position of the stud S, and through this point draw a line parallel to the central line of motion. Next move the template on the other set of $\frac{1}{2}$ stroke elements until the stud reaches the line just determined. Mark the point in which the link arc now intersects the central line of motion, and measure the distance of such inter-section from A. The required correction will be $\frac{1}{2}$ *of this quantity.* If the point falls on the forward side of the valve stroke it indicates that the exhaust chamber must be moved *bodily* this amount towards the forward edge F (Fig. 11); but if on the back, towards the back edge N.

IV. ECCENTRIC ROD PINS BACK OF LINK ARC.

Judging from the frequency with which mistakes are made in the location of the eccentric rod pins, one is apt to conclude that most Designers regard their arrangement as a mere matter of caprice, and as having little or no bearing on the symmetry of the motion. This, however, is by no means the case, for each combination has its own appropriate method of attachment. A connection of the pins *back* of the link arc is best suited to a motion like that of Fig. 22, because it tends to *hasten* the speed of the rocker pin in the forward stroke, an object usually accomplished by raising the link at the expense of slip.

The upper part of Fig. 34 has been reproduced in Fig. 40 for the purpose of explaining this peculiarity of Link No. 1. By erecting perpendiculars to the central line of motion through the cut-off points l, l', and by measuring from the same line, the respective distances of the eccentric rod pin f, for the 0.92 cut-off elements, we discover that the forward stroke distance T is much less than the return stroke R, and that if T was equal to R the rocker pin would be carried *beyond l*, thereby *delaying* the cut-off of the forward stroke which now it *hastens*. In reality the slip is avoided by elevating the eccentric rod pin instead of the link arc. The advantage gained by this feature is very great, for while it tends to equalize the motion it reduces the slip of the link and renders easy the work of suspension from a tumbling shaft.

In locomotive practice the distance of the pins' removal from the arc varies between 2¼ and 3 inches. For marine work the limits are, in general terms, double these quantities.

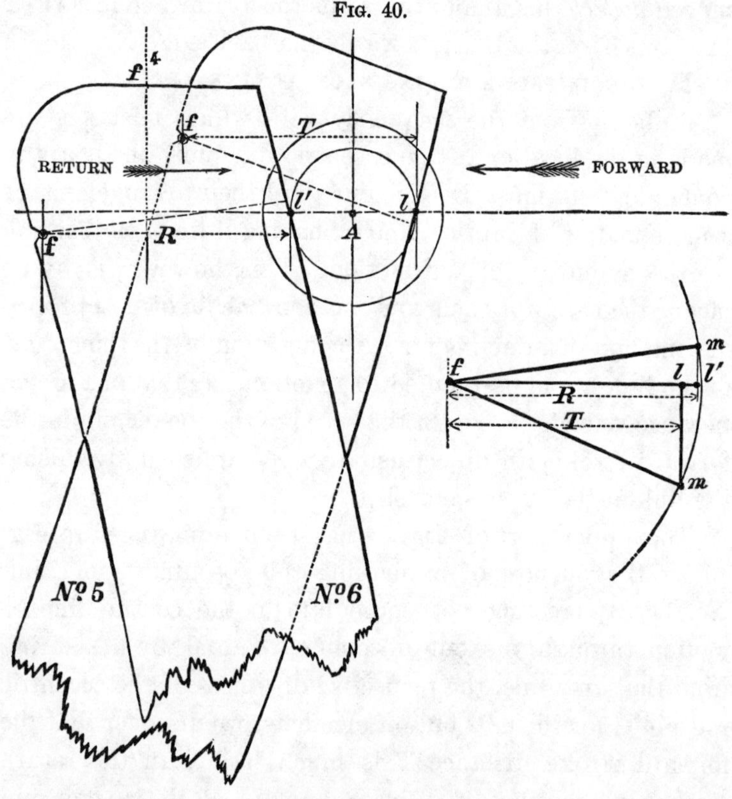

FIG. 40.

The principal fact to be borne in mind is that, within suitable limits, the *greater* this distance the *more readily* can the motion be equalized.

From the foregoing we perceive that LINK No. 1 is appropriate to a motion requiring *acceleration for any cut-off point beyond the neutral position* A, *and having its smallest crank angles laid off between the link and centre of the*

APPLICATIONS OF LINK Nº1. (Positive Motion.)

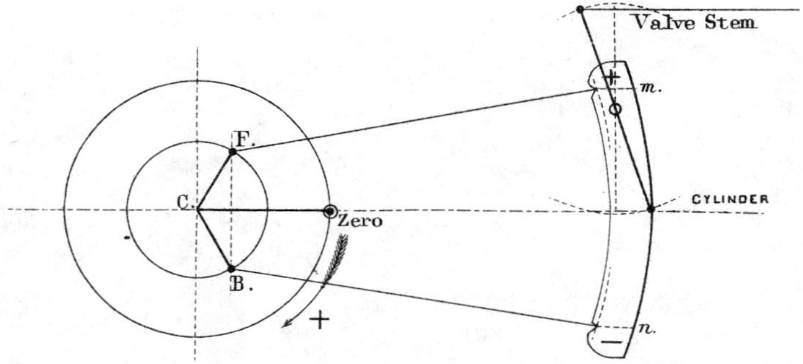

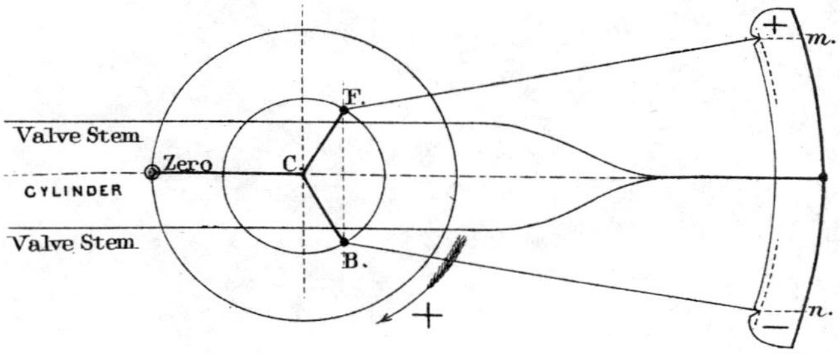

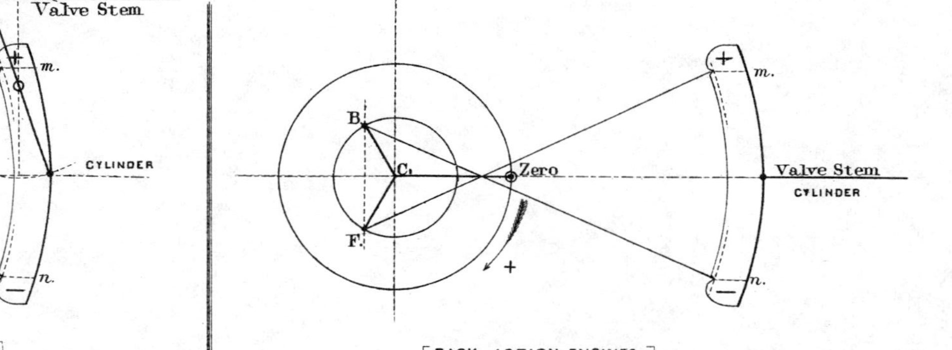

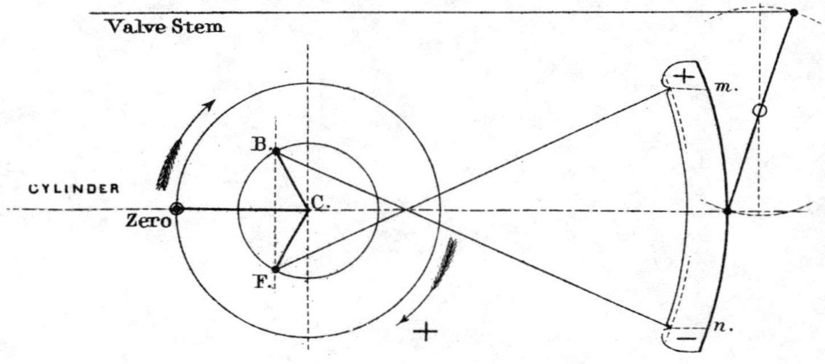

main shaft (as in Fig. 29). The four typical forms of such motions are here presented with a suitable arrangement of the parts, when the link is dropped in the full gear forward, for a positive motion of the crank. But if a negative motion be demanded, it is only necessary to transpose the entire motion so that the cylinder comes on the *opposite* side of the shaft, in other words to take the power off the opposite extremity of the shaft after first turning the engine end for end.

It was remarked (page 57) that the crank angles of a *back-action* engine are invariably the *reverse* of those common to one of direct action; hence, if we wish to preserve the identity of the motion in laying off a figure like 29 for such an engine, it will be necessary to locate the initial positions F and B of the eccentrics opposite to those proper for a direct action, and apply thereto either of the two schemes offered in the Diagram.

EXAMPLE.

The following dimensions have been taken from a most successful freight engine; their application will serve to familiarize the student with the principles of the foregoing method: 6 driving wheels, 57 inches diameter. Outside cylinders, 18 inches $\times$ 22 inches. Connecting rods 86 inches in length. Ports 15″ in length; steam, $1\frac{3}{8}$″ wide; exhaust, 3″. Diameter of eccentric circle $=5$ inches. Maximum cut-off $=0.8$ of stroke. Mid gear lead $=\frac{5}{16}$ inch. Rocker from shaft $=55\frac{1}{4}$ inches. Length of rocker arms $=9$ inches. c. to c. of eccentric pins $=13$ inches. Tumbling shaft arm $=18$ inches. Hanger $=13\frac{1}{2}$ inches. Pins back of arc $=3\frac{1}{4}$ inches.

Required.—Radius of link, distance of point of suspension back of link arc, lap of valve, full gear lead, and location of the tumbling shaft.

LINK No. II,

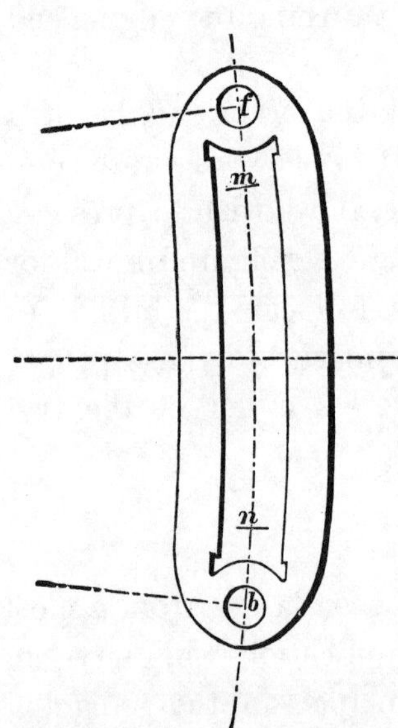

COMMONLY known as the "Open Link," is specially adapted to cases in which the link acts directly on the valve stem without the intervention of a rocker, as peculiar to British practice. It differs from No. 1 in the location of its eccentric rod pins. These, instead of occupying stations back of the link arc, reside at points f and b beyond the extreme positions m and n of the link block. They consequently move a greater distance than the latter points, and in order to preserve the same travel of valve the eccentric circle must be enlarged.

In locomotive practice the distance between eccentric rod pins f and b varies from 16″ to 20″.

The diameter of the eccentric circle varies from $5\frac{1}{2}$ to 7 inches, and the working points m and n usually about 3″ from the pins. The template for this form of link is illustrated by Fig. 41.

This link No. 2 is specially adapted to a motion requiring acceleration for any cut-off point beyond the neutral position A, and having its *greatest* crank angles laid off between the link and centre of the main shaft. It will be remembered that with Link No. 1 (Fig. 40) the *smallest* crank angles occupied such a station, and the eccentric pin was elevated to the position f, making the distance T less than R.   If, however, the *greatest* crank angles had occupied this position, the element f would have been removed

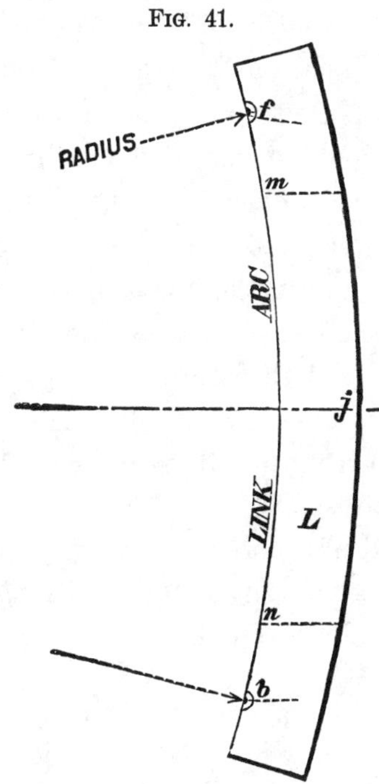

Fig. 41.

to some line f^4 nearer the shaft, while the link would have been depressed and slip increased in the endeavor to make T=R.   But the equality of these terms is established when the link arc passes through the centres of the eccentric rod pins; hence, a "Box Link," having its pins over the working points m and n, will be found better adapted to this position than Link No. 1, yet it only supplants the more readily-constructed open link, when it becomes important to retain the throw of the eccentric at its lowest limit.

As with Link No. 1, so with No. 2, there are four

schemes to which the motion is peculiarly applicable. These are given in the accompanying Diagram for the positive motion of direct and back-action Engines. As heretofore explained, their negative motion may be obtained by transposing the cylinder and link motion to the opposite side of the main shaft and taking off the power from the other extremity of the same. When, therefore, it has been decided that an engine shall have a direct or back-action arrangement of the connecting rod, the link to act through or without a rocker on its valve, and its forward motion to be positive or negative, then an appropriate arrangement of the link can be at once selected from these two diagrams, and the side of the shaft determined, on which the cylinder should be placed, to secure the desired crank motion when the link is dropped in the full gear forward.

It should be observed here, that the lower part of the link is never used in a horizontal engine to impart a forward motion because the weight of the overhanging or sustained parts tends to render the motion unsteady.

CONSTRUCTION

To find approximately the throw of the eccentrics and their angular advance for a given travel of the valve, length of link and position of the working point.

Describe about any point C (Fig. 42) on a vertical line F R a semi-circle E H with a diameter equal to the proposed travel of the valve; from C lay off C P, the extreme distance of the working point from the eccentric pin, also C S the $\frac{1}{2}$ length of the link. With S as a centre describe thé arcs C f, P m, and b R. Suppose now the cut-off must take place at 0.78 of the stroke, then from the Travel Scale

we learn that the trial angular advance should be 28°.   Lay
off C B 28° from F C and produce the line D B passing

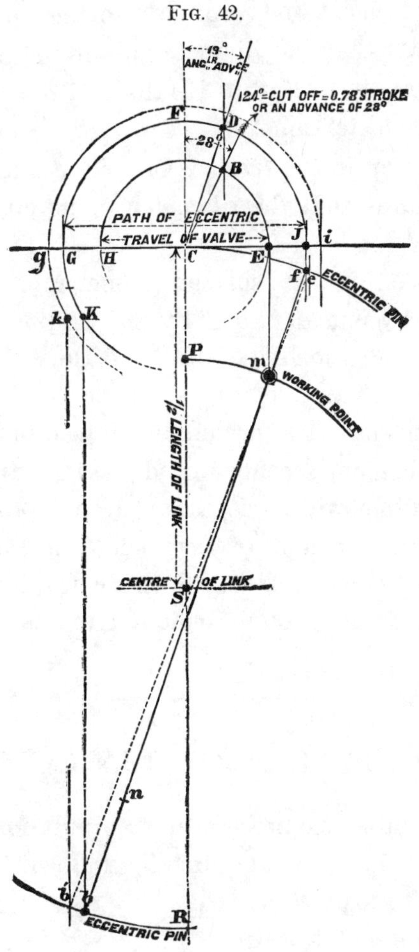

through the point of intersection B parallel to F R.   Strike
a trial eccentric circle *j* F K about the centre C.   Take the
distance between its two intersections D and F in a pair of
dividers and lay it off *twice* from G, giving the point K.

Through K draw a line parallel to C R and determine its intersection *b* with the arc R *b*. Draw a right line through *b* and the extreme travel point *m*. If now *f* should be found in the tangent line through *j* to the eccentric circle, it would prove that the diameter of this circle had been assumed correctly. But if it falls *without* the assumed circle, this diameter must be *increased*. Conversely any point *within* requires a *diminution* of the same. Thus the point *e* indicates that the diameter of its circle *j g k* has been assumed too large.

Having secured the correct diameter of the eccentric circle, join the points D and C by a right line and measure its inclination to the line F R. We thus obtain 19° the true angular advance of the eccentric for accomplishing the desired cut-off. The principle involved in this construction is that when one eccentric produces its extreme throw D (Fig. 33) the other will be separated from its like position E by the horizontal distance of T, which always equals *double* the angular advance. Although this construction does not claim strict accuracy, it will be found to answer all practical purposes.

APPLICATION OF LINK NO. II.

For illustrating the process of manipulation peculiar to this form of link, we will assume the following terms:

Crank and rod ratio $= 1 : 6\frac{1}{2}$.
Cut-off at 0.8 of the stroke.
Travel of valve $= 4\frac{3}{4}$ inches.
Valve stem from shaft $= 5$ ft.
Distance between eccentric rod pins $= 18''$.
Extreme working points, $3''$ from pins.
Mid-gear lead opening $= \frac{3}{8}$ inch.

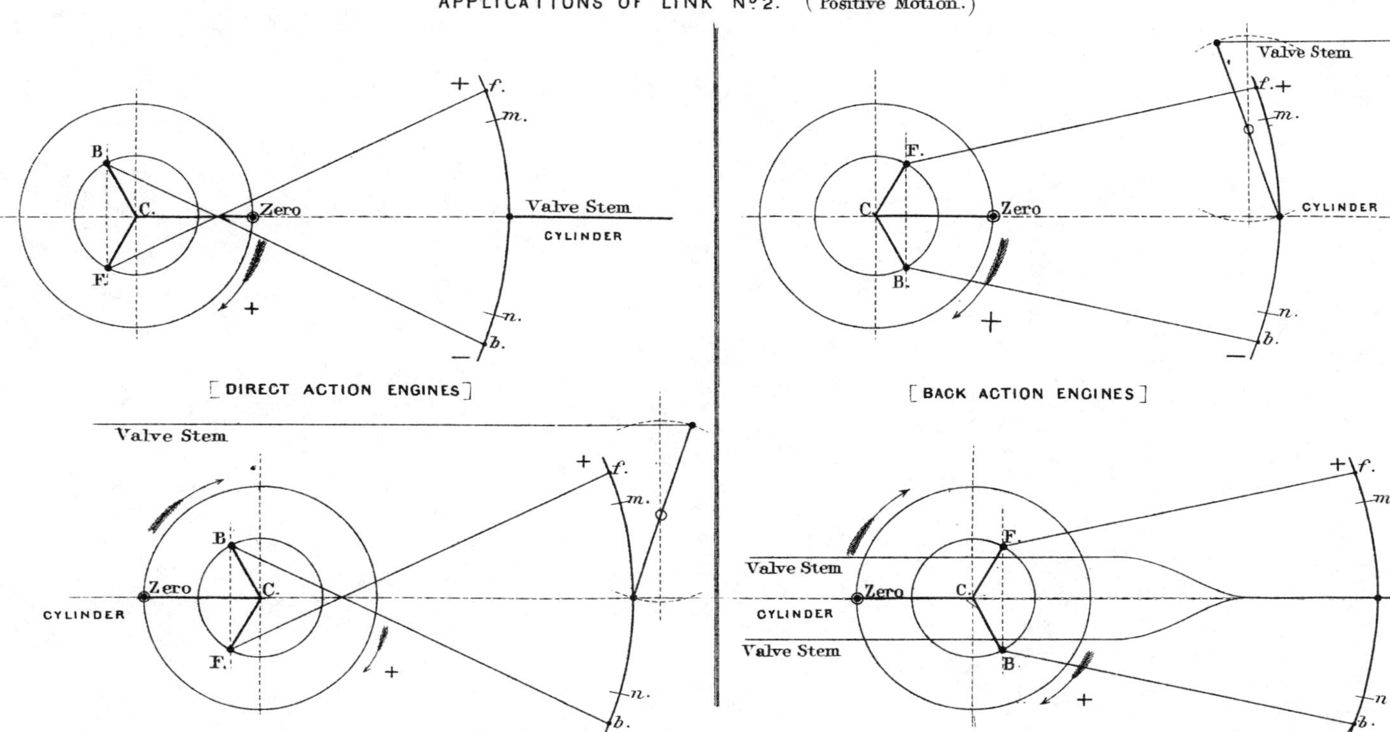

APPLICATIONS OF LINK No.2. (Positive Motion.)
[DIRECT ACTION ENGINES]
[BACK ACTION ENGINES]
CYLINDER
Valve Stem
Zero
B
F
C
b
n
m
f
+
-

By a construction like that of Fig. 42, an angular ad-vance of about 16° and eccentric circle of $6\frac{3}{4}$ inches diameter are found to conform with the above conditions. If now the connecting rod and link motion act *directly* on the piston and valve the initial position of the motion will be that pre-sented in the first Figure of the APPLICATIONS, and the four starting points F, B, f, b, together with the four $\frac{1}{2}$ stroke points of the eccentric circle, may be mapped in a similar manner to Fig. 29 ; while the four 0.8 cut-off points should be laid off by means of a protractor from the lines F C, B C, f C, and b C.

Having mapped these twelve important positions and constructed a template like Fig. 41 the next step will be to follow the journeyings of the link arc and locate the tum-bling shaft in a manner precisely analogous to that pursued with Link No. 1 :

Find $\left\{\begin{array}{l} \text{1st. The mid-gear travel} = d'\ d^2. \\ \text{2d. The lap A } l \text{ for a given mid-gear lead.} \\ \text{3d. The position of the stud for equal } \frac{1}{2} \text{ stroke} \\ \quad\quad \text{cut-offs of the piston.} \end{array}\right.$

This form of link is more generally suspended from the upper eccentric rod pin than from a point midway between the two, and has the tumbling shaft *below* the central line of motion ; but by marking the various locations of both pins, as well as the centre line j, we will be able to select the most appropriate of the three suspensions.

Let us follow first the central suspension. We find that by locating the stud on the arc, the line $c\ c$ of its motion, for the $\frac{1}{2}$ stroke cut-off, becomes parallel to the central line of motion, a condition suited to ready suspension. But on

dropping the template to the full gear cut-offs the stud assumes the positions S^4 S^5, Fig. 43, whose line c^2 c^2 has so
great an inclination to the central line of motion, that it
would be next to impossible to successfully hang it from a
tumbling shaft and accomplish *equal* cut-offs in *all* gears.
We must consequently resort to a change of link arc radius
and unequal full gear leads (as in Figs. 38 and 39) in order
to establish perfect equality.   To do this, raise the link arc
so that the stud shall occupy a location S^6* thereby bringing
c^2 c^2 more nearly to a state of parallelism with the central
line of motion.   Mark l^2 the new lap point; strike a new
lap circle l l^2 with neutral position a, *nearer* to the main
shaft, and about this point describe the mid-gear travel circle giving new points d^4 d^5 through which the link arc must
pass for the mid gear.   Bring the template to position No.
2, as in Fig. 31, mark the eccentric pin points f and b; then
search for a radius whose arc shall pass through the three
points f, d^5, b.   This radius will always prove *greater* than
the distance from the centre of the shaft to the neutral position A instead of *less*, as found with Link No. 1.

Having obtained a correct link arc, cut out a new template like Fig. 41 to represent it, and reconstruct the entire
motion.   The new radius will equal 5 ft. 7″.   The corrected
lap$=1\frac{1}{32}$ inches.   The lead for the forward stroke, increases
from $\frac{1}{8}$ to $\frac{3}{8}$ in the mid gear, and from $\frac{3}{16}$ to $\frac{3}{8}$ for the return
stroke.   The lines of suspension pin motion c, c', c^2, c^3 may
converge, as in the present instance, towards some point
beyond the link.   In such case the tumbling shaft should
be placed on the side of the convergence.

If, however, the link is suspended by the upper eccentric rod pin, the forward gear paths of the suspension pin
motion will be represented by the lines h, h', and the tumbling shaft will lie below the central line of motion; but for

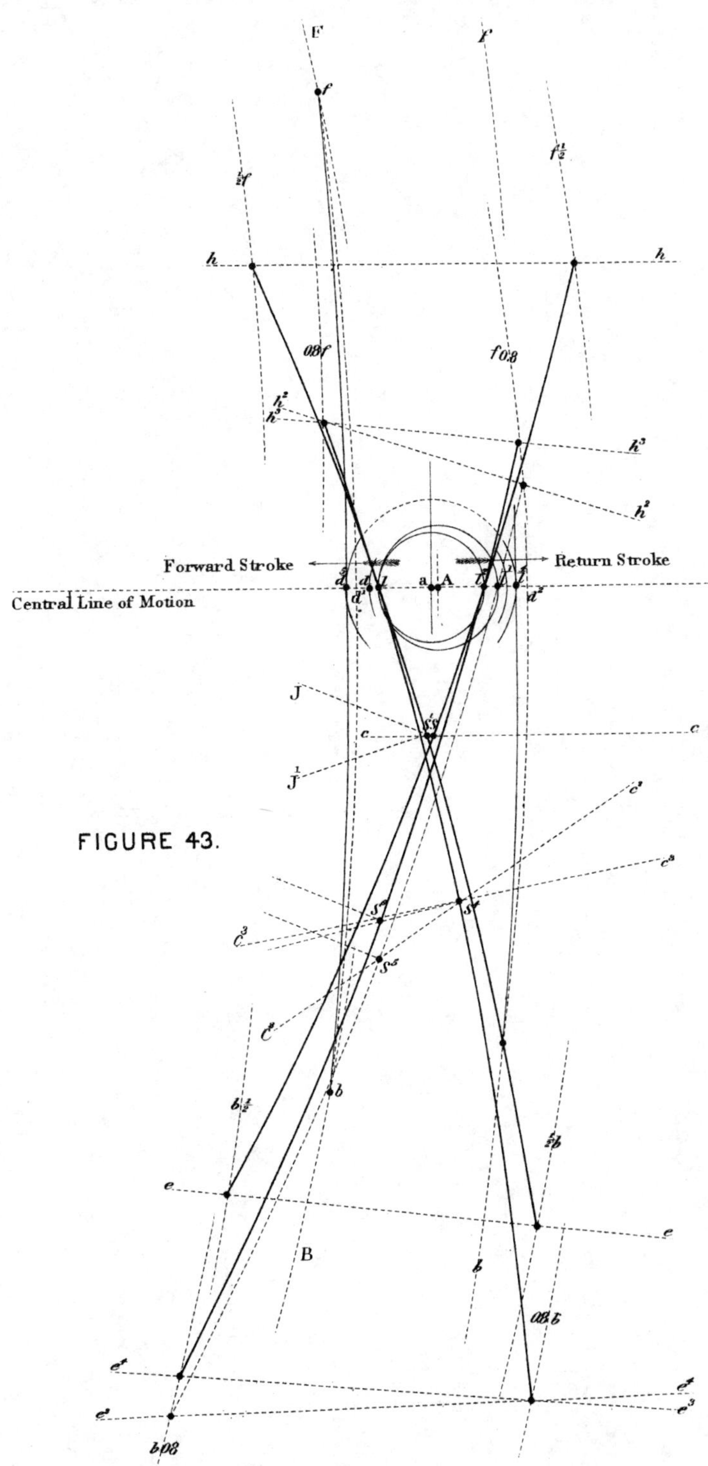

F
F'
f
f½
½f
h
h
080f
f080
h²
h³
h³
h³
h²
Forward Stroke
Return Stroke
d
d'
d
d²
a
A
d²
Central Line of Motion
J
ss
c
c
J
c'
J
c³
FIGURE 43.
s²
s'
C³
c²
s⁵
C²
b½
½b
b
e
e
B
b
080b
e'
e'
e³
e³
080b

the lower eccentric rod pin, the lines will become e, e^4, e^3, and its tumbling shaft will stand above this line.

The judgment of the designer will in every case decide to what extent the lead openings should vary in the full gears, what inequalities of cut-off are admissible, and whether or not the motion should be equalized simply for the forward without regard to the back gear.

While attempting to analyze the peculiarities of various existing link motions, the investigator usually finds that among other dimensions, only the lap and mid gear lead are given, and that no allusion is made to the angular advance, or to the maximum cut-off. To solve such cases, he is compelled to work the problem as it were, *backwards*. On the central line of motion he plots the lap and lead, as on Fig. 32, places the template in the Nos. 1 and 2 positions, as in Fig. 31, then with the length of the eccentric rod as a radius and the positions of the eccentric pins as centres, strikes in the direction of the shaft indefinite arcs, whose intersections with the eccentric circle give the points F, B, f, b, (equi-distant from the central line of motion) and from these the desired angular advance can readily be measured.

BOX LINK.

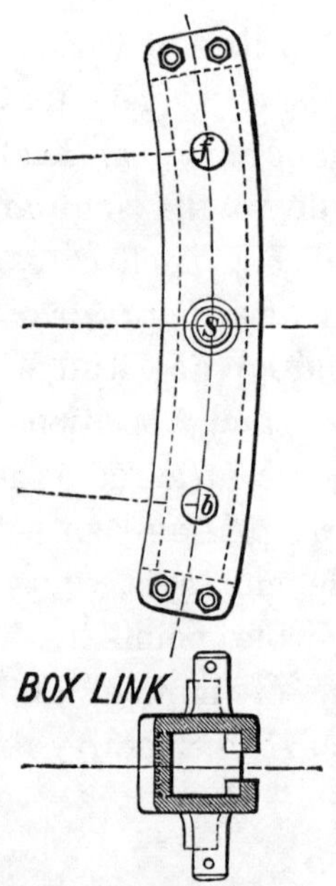

ALTHOUGH the mechanical construction of this form of link is rather more difficult than that of No. 2, it serves the part of a good substitute when a short throw of the eccentric is required ; for with it the maximum travel of the valve always approximately equals the diameter of the eccentric circle.

At times the box link can be employed in positions appropriate to Link No. 1, but *very rarely* with those *good results* in respect to minimum slip which obtain with the former motion. On such occasions the stud usually lies at some point *beyond* the link arc, determined by placing the link in the $\frac{1}{2}$ stroke cut-off positions, Nos. 3 and 4, and plotting the centre line *j* as in Fig. 32.

When, however, the box link is used in place of Link No. 2, the centre of suspension S generally falls *within* the link arc, or between it and the main shaft.

The construction of these links varies ; in some instances the ribs are formed on the inside as represented, while in others, they are cast on the sliding block and overlap the link plates.

STATIONARY LINK.

THIS form of connection between the valve and eccentrics is specially applicable to those circumstances in which the former requires no rocker. The mutual relation of the parts will be clearly perceived from an examination of Fig. 44 which illustrates one of the most successful methods of suspension.* The eccentrics stand in their usual location for a direct action motion. The main link is hung from a *fixed* point by a short bar called the " suspending link" and the link block connected with the valve stem through the " Radius Rod" $m\,d'$. By means of a reversing combination the block may be carried to any point between m the full gear forward and n the full gear back. But since the link arc is always struck with a radius equal to the length of the rod $m\,d'$, having its centre at d' and d^2 in the central line of motion, when the crank occupies the zero or 180° location, it must be evident that the block can be moved from one full gear to the other *without* altering the position of the points d' or d^2, consequently the lead opening will remain *constant* throughout the motion, the same as in Fig. 17. Now it has been invariably the custom to simply define a stationary link motion as *"one in which the lead is constant"*

* For convenience of observation, the cross sections of the valve and seat have been revolved to a plane at right angles to their true position.

leaving it to be inferred that the angular withdrawal of the crank from its zero position at the moment of pre-admission must also be a *constant* quantity, whereas in reality this lead angle *increases* just as much for a stationary link motion as for a shifting one. The only difference between the two is that the lead opening of the stationary link motion is more ample and the angle slightly greater, for all except the mid gear, than with the shifting link motion. But this distinction has been so clearly drawn in Part III. that further remark can scarcely be necessary. Unlike the shifting link motion, however, the lead opening is *not* dependent on the arrangement of the eccentric rods, for these may either be crossed or opened without altering the result. But for the purpose of meeting the other conditions of the motion an arrangement like Fig. 44 should be adopted.

As a general thing more attention is paid to the equalization of the cut-off and reduction of the slip in the forward than in the back gear. For the accomplishment of this object, the centre of the link should be dropped *below* the central line of motion, the angular advance of the backing eccentric slightly reduced and the backing eccentric rod lengthened.

The simplest method by which the student can obtain a clear idea of the action of the parts, in a stationary link motion, will be for him to take the dimensions of some successful motion, cut out a proper template for the link and trace its journeyings throughout the different gears in conformity with principles already laid down for the shifting link motion.

The following dimensions (in absence of others) will answer such a purpose :

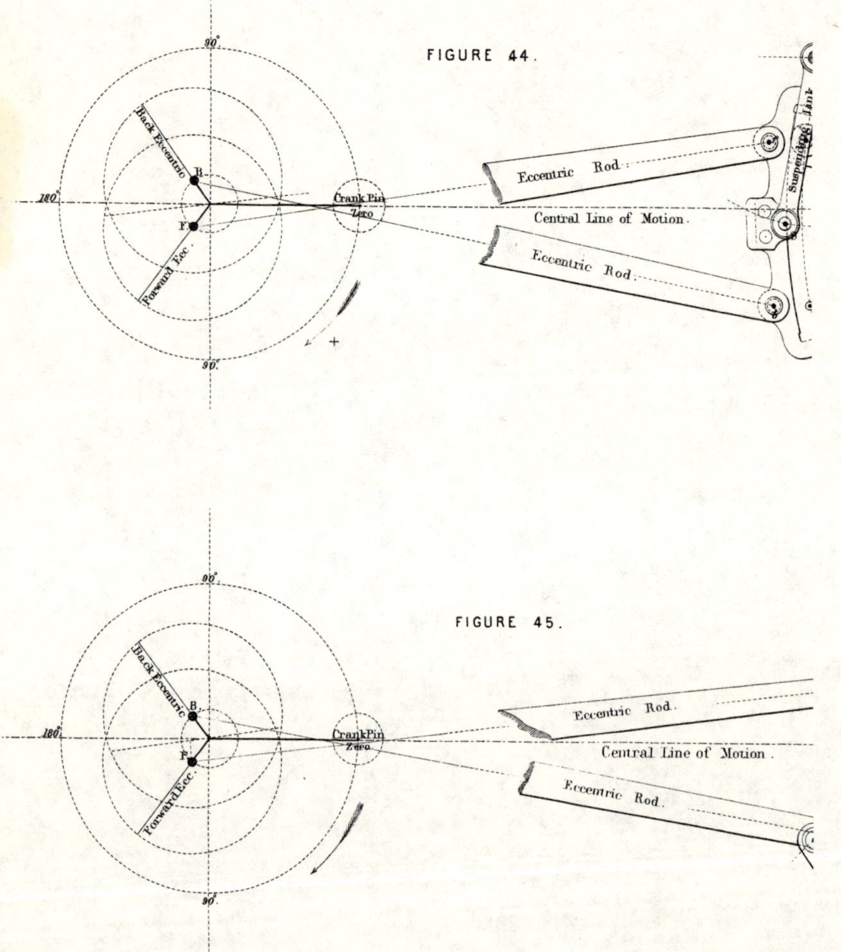

FIGURE 44.
90°
Back Eccentric
B.
180°
F.
Forward Ecc.
90°
+
Crank Pin
Zero
Eccentric Rod.
Central Line of Motion.
Eccentric Rod.
Suspending Link

FIGURE 45.
90°
Back Eccentric
B.
180°
F.
Forward Ecc.
90°
Crank Pin
Zero
Eccentric Rod.
Central Line of Motion.
Eccentric Rod.

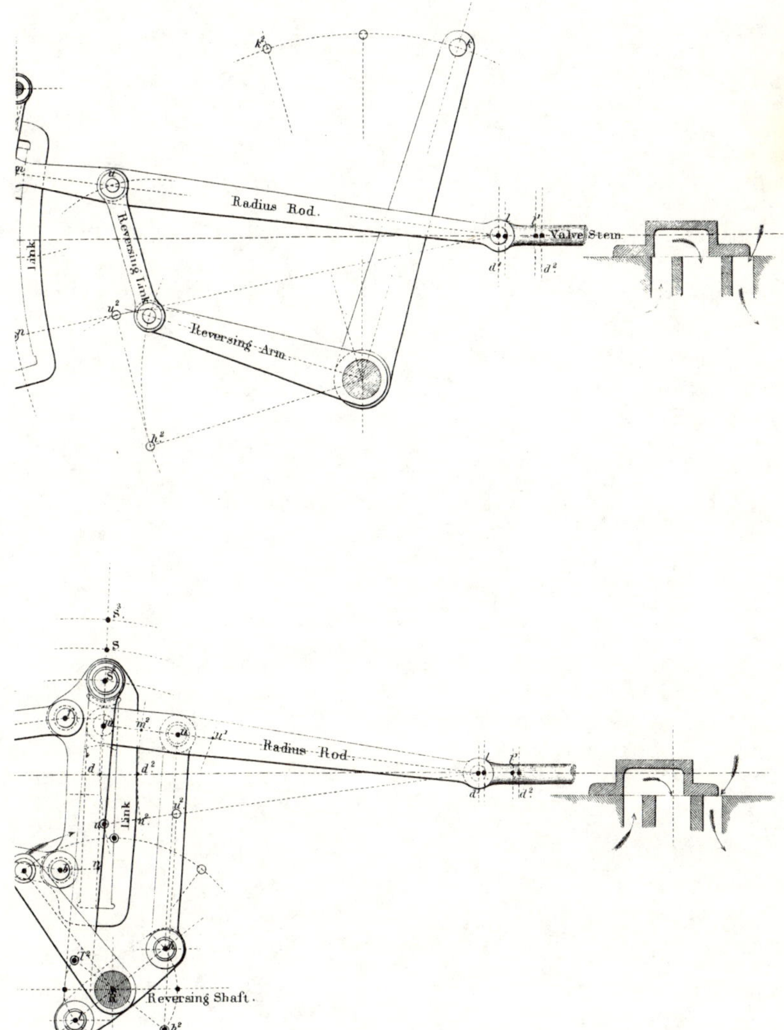

Radius Rod.
Reversing Link.
Link.
Reversing Arm.
Valve Stem
Radius Rod.
Link
Reversing Shaft.

Diameter of piston$=18$ inches.
Stroke$=24''$; Connecting rod$=91''$.
Ratio$=1:7\frac{1}{2}$.   Throw of eccentrics$=2\frac{5}{8}''$.
Forward eccentric angular adv.$=27\frac{1}{2}°$.   Rod$=57\frac{3}{8}''$.
Backing eccentric angular adv.$=26°$.   Rod$=58''$.
Eccentric rod pins $12\frac{1}{2}''$ apart, $3''$ back of arc.
Centre of suspension $1\frac{1}{2}''$ back of arc, $1\frac{1}{8}''$ below line.
Radius rod$=37''$, Reversing link$=11\frac{1}{2}''$, Hanger$=9''$.
Tumbling shaft arm$=18''$.   Reversing pin $8''$ back of arc.
Lead$=\frac{3}{8}''$, Steam port $=2''$, Exhaust$=3\frac{1}{2}''$.
Maximum travel about $4\frac{3}{4}$ inches.

The Stationary Link is seldom found in American practice from the fact that all modern locomotives are built with steam chests on top of their cylinders, instead of at the side. On stationary engines, the link and governor are occasionally used conjointly; in such instances the stationary link will be found best adapted to the requirements of the case, because its radius rod imposes a far lighter duty upon the balls of the governor, than the shifting link with its rods, hanger, and additional friction of eccentric straps.

ALLAN LINK MOTION.

THE discovery of this motion was a natural sequence to the invention of the shifting and stationary links. By it a compromise has been effected between the leading features of both motions, resulting in a more *direct* action and perfect balance of the parts together with a reduced slip of the link block. One mode of suspension, for the link in the full gear forward, appears in Fig. 45, in which the cross sections of the valve and its seat have been revolved for the purpose of more plainly exhibiting their relative positions. The locations of the point of suspension and attachments of the eccentric rod pins upon or back of the link arc are quite as variable for this, as for the shifting link motion ; and the requirements of the other details generally indicate whether the reversing shaft should be placed above or below the central line of motion.

In proportioning the parts, the main object is to move the link and radius rod (when the crank stands at the zero or 180° locations) in such a manner that the link arcs peculiar to each motion shall always be *tangent* to each other. In this case all the locations of the link block will be found in one and the same *straight* line. This peculiarity has given rise to the title "Straight Link" motion expressive of the form of the link.

The radius rod and main link are supported by rods from the reversing shaft arms, and the inequality in the lengths of the latter, which is essential to a proper suspension of the parts, incidentally tends to *equalize* the weights resting on the opposite sides of the reversing shaft, thus greatly facilitating a change of the motion from one full gear to the other.

Well-schemed motions of this type practically preserve the characteristic feature of the stationary link, viz., a constant lead ; yet from the nature of the case they possess at times slight inequalities in one or both of the full gears. These, however, are quite insignificant for a relatively long radius rod and short travel.

The ratio between the long and short arms of the reversing shaft may be readily determined for any given travel, angular advance, length of eccentric rods, link and radius rod, by placing the template in the No. 1 and 2 positions (Fig. 31), marking the mid-gear travel d' d^2, sweeping indefinite arcs through these points with the radius rod, and drawing down the template, or centre of suspension, from S to S^2 until its straight line intersects these arcs in points m, m^2. Then map the radius rod m d', m^2 d^2 giving the points u, u' of the reversing rod pin above the central line of motion. The centre l of the link arm pin must fall as much below the horizontal line through the reversing shaft R as S^2 does below S. In like manner the pin h of the other arm must rise as far above the horizontal as u does above the central line of motion. Finally, draw l h of length sufficient to accommodate the details of the shaft, and we have at once the proper dimensions for the reversing shaft arms.

It should be observed that the tendency of m^2 is to drop below m and thus distort the motion, but this will be obviated if the two are brought into one horizontal line *inter-*

mediate between such stations.   The change will result in a *slightly* increasing lead, as is common with the shifting link motion.

The following dimensions can be employed for an investigation of this class of motions:

Diameter of cylinder=16″.

Stroke=24″ ; Connecting rod=87″ ; Ratio=1 : 7¼.

Throw of eccentric=2¼″ ; Angular advance=26°.

Eccentric rods=39½ inches.

Radius rod=47″, connected 7″ back of link.

Box link, suspended by stud at centre with eccentric rod pins 10″ apart.

Suspending rods both 18′ long.

Reversing lever, long arm=6″.

Reversing lever, short arm=2½″.

Mid-gear lead=¼ inch.

Steam port=1¼″ ; Exhaust=2¾″.

If the examiner desires to analyze any stationary or straight link motion already constructed, having a given lap and lead but no specified angular advance, he should work the problem *backwards*, as follows : First locate the four cardinal points d', l, l', d^2, and sweep their arcs with the radius rod ; then place the link in the positions Nos. 1 and 2, with the positions of the eccentric rod pins as centres and the length of the eccentric rod as a radius, sweep the four arcs, which must contain the points F, B, f, b, and describe through them the given eccentric circle, in such a manner that all the points of intersection will lie equally remote from the central line of motion.   With these initial points determined, the investigation can proceed on the principles already explained.

WALSCHÄERT LINK MOTION.

This device groups two perfectly distinct motions—the one derived from a single eccentric, the other from the cross-head of the piston rod—in such a manner that their combined effect is, when the parts are well proportioned, quite analogous to the motion obtained from the stationary link. From the nature of the connection between the cross-head and the valve-stem, the motion can be more readily applied to an outside cylinder engine than to an inside one.

The eccentric usually assumes the form of a return crank from the main crank pin, as shown in Fig. 46. Its centre then is found on a line at right angles to the crank arm. The angular advance becomes equal to zero, and hence, so far as the link will be concerned, the valve can have neither lap nor lead. The link oscillates freely about a fixed axis, and its arc has a radius equal to the length of the radius rod. This rod is moved from one full gear to the other, in the usual manner, by means of a reversing shaft with arms, From the extremity of a short arm, rigidly bolted to the cross-head pin, extends a union bar which is pinned to one end of the combination lever. By the aid of this lever, the eccentric and cross-head motions are so combined, that the latter virtually restores the angular advance discarded while locating the eccentric, and consequently enables the valve to possess both a constant lap and lead.

The truth of this assertion will appear from an examination of the lever elements (Fig. 47) for 12 locations of the crank arm.

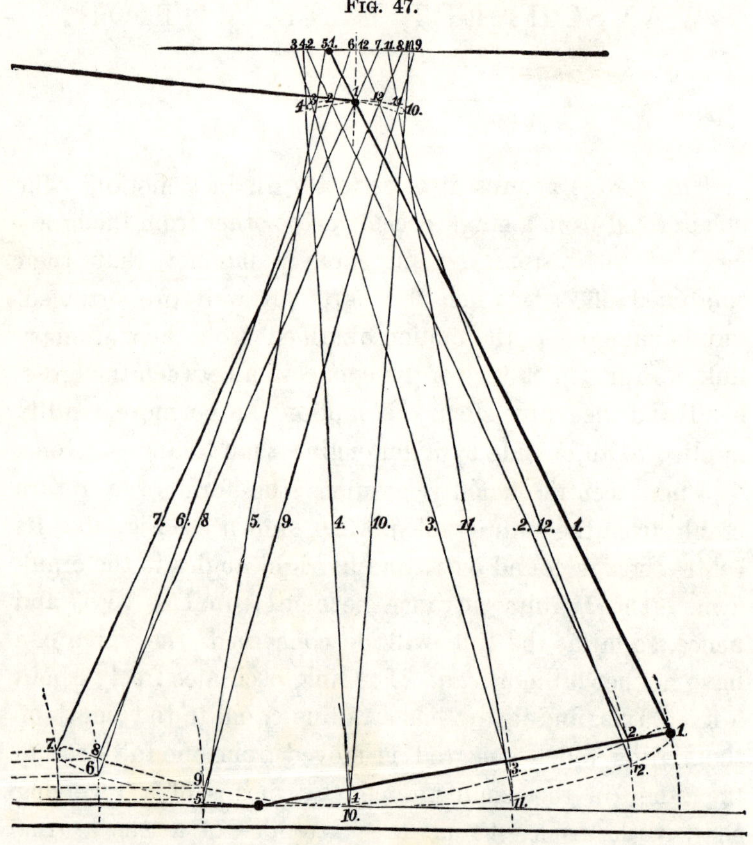

FIG. 47.

The subjoined dimensions will be found convenient for the construction of a trial example, in which of course two templates should be employed, one to represent the link, the other the combination lever :—

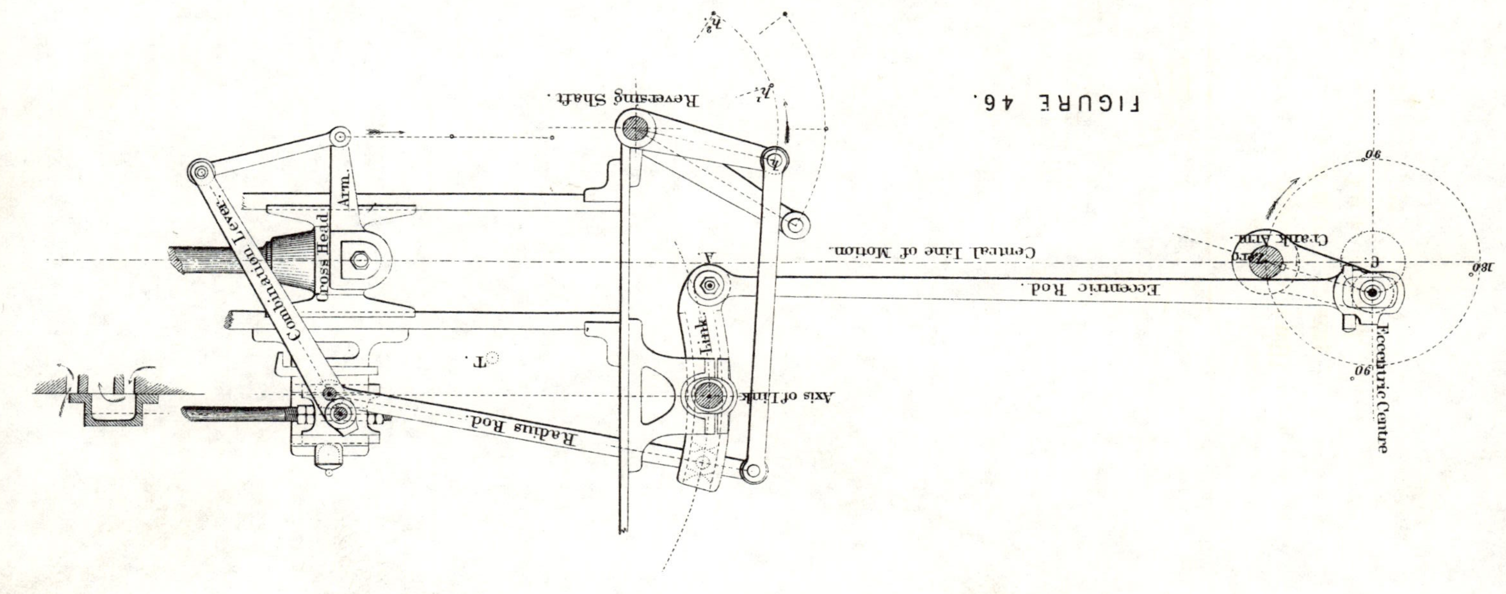

Combination Lever.
Cross Head.
Arm.
Radius Rod.
T.
Reversing Shaft.
Axis of Link.
Link.
A.
Central Line of Motion.
Eccentric Rod.
Crank Arm.
Eccentric Centre.
90°
90°
180°
h_1^1
h_2^2
FIGURE 46.

Diameter of piston=18″.

Stroke=24″ ; Connecting rod=100″.

Ratio=1 : 8⅓.

Axis of link 74″ from centre of shaft and 15¾ above central line of motion.

Eccentric rod pin 13″ from axis of link.

Radius of link and length of rod=42½″.

Link block to end centre of radius rod=6″.

Long lever of combination arm=30″.

Cross-head arm drops 14¾ inches.

Connection for arm and lever=16″.

Throw of eccentric=3½ inches.

Travel of valve (maximum)=4½ inches.

Lap=1 inch ; Constant lead of ¼″.

Steam port 1¼″ ; Exhaust port=3″.

Cases will at times arise in which the distance between the centre of shaft and valve, will prove too contracted for a satisfactory arrangement of the parts after the manner shown in Fig. 46. In such instances the curvature of the link should be reversed, the radius rod made to lie between the link and shaft, and the valve stem lengthened, to adapt it to the new position of the combination lever.

The Designer will find that his efforts, towards the equalization of the cut-off in the Walschäert Motion, are attended with far less difficulty than similar ones with the shifting and stationary links. This peculiarity arises, from the intimate relation constantly maintained between the valve and piston motions, through the medium of the combination lever. In consequence of which, any undue acceleration or retardation of the piston motion is immediately accompanied by like effect in that of the valve, thereby greatly diminishing its capacity to derange the events of the stroke.

PART V.

INDEPENDENT CUT-OFF,

CLEARANCE, ETC.

INDEPENDENT CUT-OFF.

It has been demonstrated in Part I. that a very early cut-off is incompatible with the economic action of a single eccentric, for beyond the limit of about $\frac{2}{3}$ of the stroke, the compression attending an earlier closure of the exhaust will usually furnish a resistance in *excess* of that required for neutralizing the momentum of the reciprocating parts. If therefore a more extended range of the cut-off should be demanded, for comparatively slow speed engines, other means must be sought by which it can be controlled without affecting the exhaust; in a word *independent* valves must be introduced having a motion different from that of the main valve. For this purpose the parts are usually arranged as shown in Fig. 48. The valve A carries on its back two cut-off valves B B' and rigidly supports by pillars the surface plate H H on which a brass packing ring F F bears, enclosing a space D in communication with the condenser through the pipe E. The partial vacuum formed in this space relieves the valve in great measure of the immense pressure exerted by the steam, and consequently reduces the friction as well as facilitates the starting of the engine.

On the cut-off valve stem are turned a right and left hand thread, so that by revolving the same, the valves may be drawn closer together or separated by a wider distance according to the requirements of the cut-off. The main valve A has a lap, lead and exhaust closure appropriate to the value of the maximum cut-off, and permanently retains

Fig. 48.

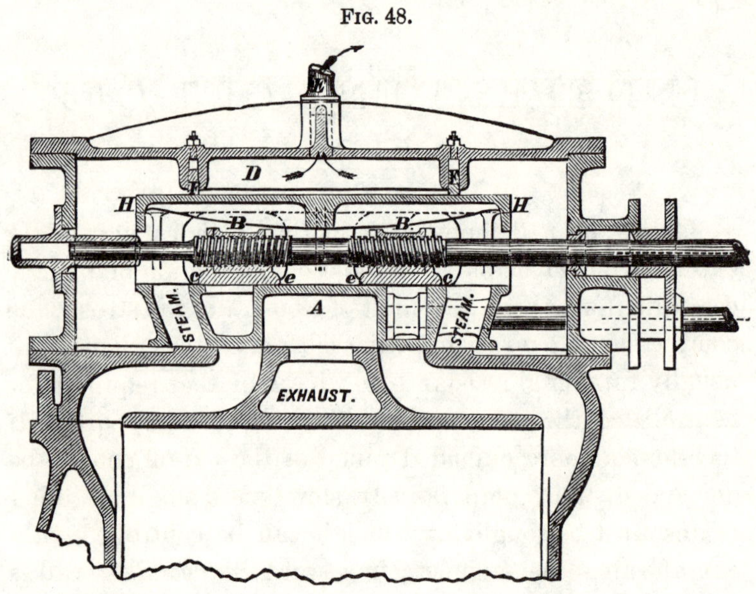

these quantities throughout every variation in the point of cut-off brought about by the separation of the valves B B'.

Its general design depends upon the range of the cut-off. For one varying between zero and about 0.6 of the stroke the combination shown in Fig. 48 is most suitable, where the *outer* edges $c\,c$, effect the closure of the port, and the valves have a motion directly the *reverse* of that peculiar to the piston. For cut-offs varying between 0.3 and 0.9 the stroke the *inner* edges $e\,e$ should be made to perform this office, and the valves should have a travel *coincident* with that of the piston. Between the assigned limits such valves give sharp and decisive cut-offs for a proper ratio between the travels of the main and cut-off valves. When the *outer* edges are employed the travel of the cut-off valves should *exceed* that of the main valve ; but with the inner edges it may *equal* or exceed it according to the degree of rapidity desired in the action.

The relative motions of the valves, may for any required case be conveniently examined, by a method indicated in Fig. 49, and the travel regulated to meet the conditions of a good motion. The valves are here supposed to cut off with their *inner* edges but the process is equally applicable to the opposite relation. 1st. Stretch a sheet of paper and upon it draw the steam ports, exhaust port, bridges, and a circle *f g* representing the motion of the main valve's eccentric with an angular advance position R 1, also the ½ stroke R 3, and the maximum cut-off R 4. Project these points to the line of the valve seat, thus giving the positions 1, 2, 3 and 4. Secure a slip of paper X A with its *base line* A over the point 6, and draw the main valve with lap appropriate to the maximum cut-off precisely as in Part I. Make the steam passages S S convergent in order to economize space. Secure a second slip W D as before shown, and above its base line D describe a circle *h j* to represent the path of the eccentric acting on the cut-off valves. Its initial position will be found at R 1 the *same* relatively as the crank of the engine. From this lay off R 2, 3 and 4, making angles with R 1 equal to those found in the circle *f g*. These preliminary steps completed, our object will be to find the degree of separation required between the valves for the given cut-offs, their width, and also to what extent the faces L L should be lengthened to prevent a fall of either cut-off valve. Suppose for instance the limits were ½ and ⅞ stroke, we *first* loosen the two strips W and X, place their base lines A and D at Nos. 3 the ½ stroke locations for their respective eccentrics, and mark e for the point at which the cut-off valve should stand to effect the closure of the steam port. *Second*, move the slips W and X until their base lines A and D correspond with Nos. 4, the other cut-off limit mark e'' and it will appear that the valve stem must be rotated, until the cut-off valve B moves a distance n from its first position e. 'Since the right and left hand threads have one pitch, the other cut-off valve B' will be moved a like distance from the common centre C. *Third*, the valves must be sufficiently wide to guard against a *re-opening* of the port before its final closure by the main valve. This distance may be found by placing the strips at the Nos. 4 positions and marking a point c on the strip W opposite the edge d ; when the length of the valve should at least equal the distance from e to c. *Fourth*, Place the strips W and X at the Nos. 1 locations and we obtain substantially the extreme position of the cut-off valve with reference to the main valve which will indicate the proper extension for the faces L L.

If other positions of the eccentric are interpolated between 1, 2, 3, and 4, the relative motions may be accurately traced and the degree of port opening observed, so that in event of the latter proving inadequate, the diameter of the circle $h\,j$ can be increased a suitable amount. It is customary in proportioning stationary engines with valves constructed on this principle, to make the cut-off variable between 0.3 and $\frac{3}{4}$ the stroke, while the main valve is arranged with a lead angle of about 8° and a lap suitable to a cut-off of $\frac{7}{8}$ the stroke. The resulting angular advance usually furnishes an appropriate exhaust closure. If, however, on trial the crank should not pass its centres smoothly, the angular advance of the eccentric must be increased or diminished until the proper compression is discovered for counteracting the momentum of the reciprocating parts, as

FIG. 49.

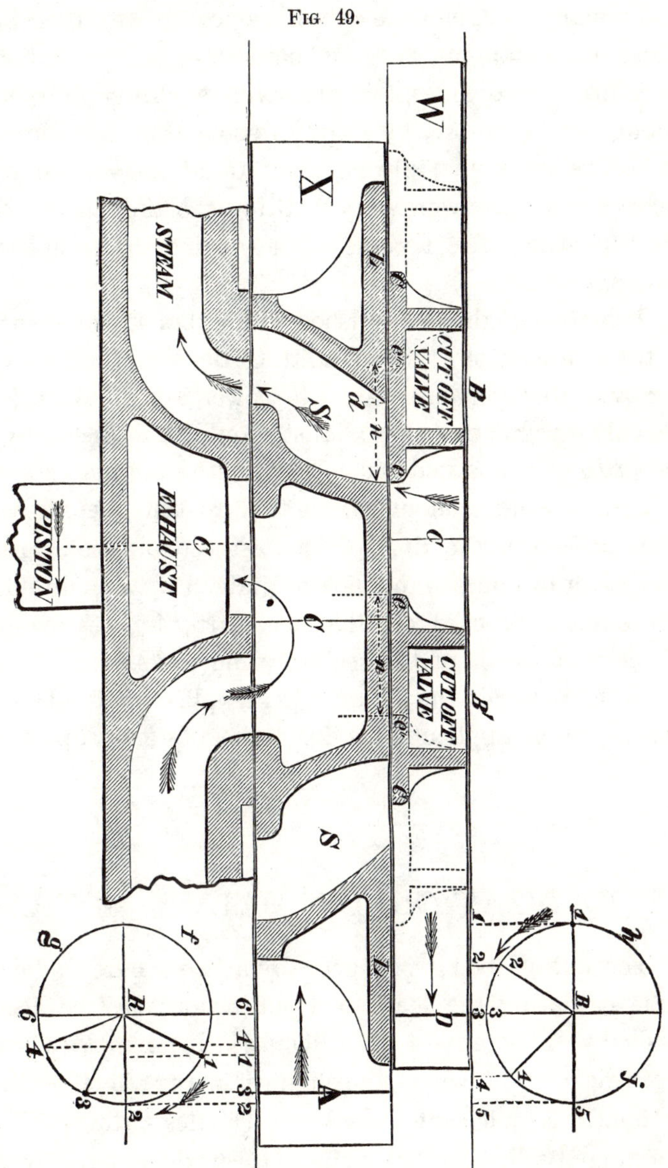

well as lead opening for neutralizing the effect of lost
motion.

In marine engines the cut-off valves act with their *outer* edges. The lap angle of the main valve is then taken at from 15° to 20 degrees, and the exhaust closure effected at about 0.9 the stroke, by simply raising the link, until the eccentrics have virtually an angular advance of 35 to 45 degrees; in other words, by working the link in less than the full gear. The necessity of adjusting the eccentrics is thus obviated.

When an engine is furnished with exhaust passages perfectly distinct from those admitting the steam to the cylinder, and the demands upon its power are quite uniform, the valves regulating the exhaust should be adjusted by the *quantity of coal* consumed. That is, the angular advance of their eccentric should, from time to time, be increased until at length the limit of greatest economy is attained. The result of course must not be judged by a computation of the indicator card, for that necessarily ignores the effect of the momentum of the reciprocating parts, since it measures the power in the *act* of being applied and not *subsequently* to the application, a distinction of great importance.

EQUALIZATION OF VALVE MOTION.

The cut-off being produced by a valve independent of that regulating the lead, its equalization may be accomplished without any of the difficulties incident to a single eccentric motion, where the slightest change in either event is immediately felt by the other. The desired result is approximately obtained for a single eccentric motion, by simply *lengthening* the cut-off valve stem an amount dependent on the ratio existing between crank and connecting rod.

If, however, the valve moves under the influence of a lever attached to the reciprocating parts of the engine, as in Fig. 52, the motion will become equalized by virtue of the irregularities thus introduced to the valve motion (a counterpart of those peculiar to the piston) and any slight inequalities of the main valve will be counteracted by a change in the length of the cut-off valve stem. The lead should be made *equal* for both faces of the main valve, and if neither the single eccentric nor link by which the latter is operated accomplishes an equalization of the exhaust closure, then a suitable amount of inside lap and clearance (as explained Part II) should be given to the exhaust edges of the valve face. The cut-off equality of the main valve, in itself considered, is of little consequence, hence the link should be arranged to give the *smallest* amount of slip attainable with an equality of the lead, and the cut-off should be regulated solely by the cut-off valves.

CLEARANCE.

This term is used to express, the extent of the space which exists between the piston at the extremity of its stroke and the valve face, or the cubic contents of the steam passage plus the unoccupied portion of the cylinder. Since for each stroke of the piston this space must be filled with steam, which in no way tends to improve the action of the parts, but rather increases the amount to be exhausted on the return stroke, it becomes desirable in long-stroke engines to locate the valve face as near the end of the stroke as possible, thus reducing the cubic contents of the passage and by its directness admitting the steam with a higher initial pressure than could be obtained through a more tortuous channel.

A convenient method for accomplishing this result, is to separate the valve faces F and N by means of a stem, as shown in Fig. 50, and forming them either in the shape of a

FIG. 50.

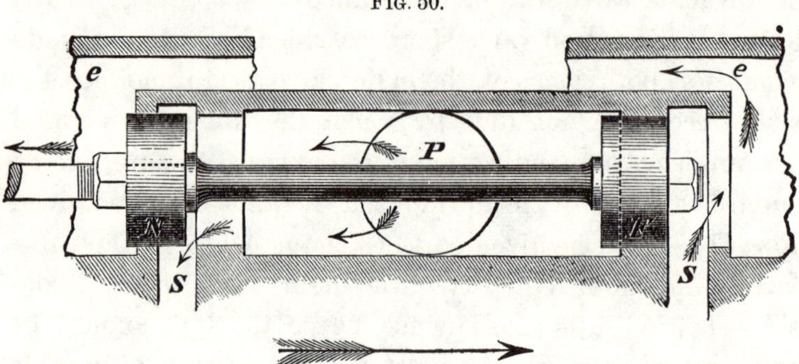

letter ▢ laid on its side, or as small pistons. The former or D valve construction was much employed a few years since, but has been gradually supplanted by the more mechanical arrangement of the piston valve. This, instead of being surrounded by packing in the valve case, carries its expansive packing in the same manner as an ordinary piston. The steam is received either between the pistons, from a pipe P, giving the most direct admission to the cylinder, or externally from chambers $e\,e$. The parts surrounding the valves are well jacketed to prevent unequal expansion, and slight irregularities are compensated by the elasticity of the packing.

FRICTION.

The friction of a valve is quite independent of its surface, except so far as the latter may increase the area upon which the steam can exert an unbalanced pressure.

There are various expedients for relieving such pressure. Two of these have been illustrated in Figs. 48 and 50. A third consists in casting a standard in the exhaust port on which is bolted a concave plate, scraped to a bearing upon the inner surface of the valve, and holes drilled through the top admit the steam *under* the valve, which tends to relieve the external pressure. Besides these devices the valve is frequently mounted on steel rollers about $1\frac{3}{4}$ inches diameter resting on plates of the same metal, and thus the sliding converted into rolling friction. Heavy valves standing in a vertical position have additional rollers under their lower edge for supporting their great weight.

REDUCTION IN TRAVEL.

The work expended in friction depends directly upon the distance traversed by the valve. Hence, in marine engines every means possible should be employed to reduce this quantity to its minimum value. This object may be

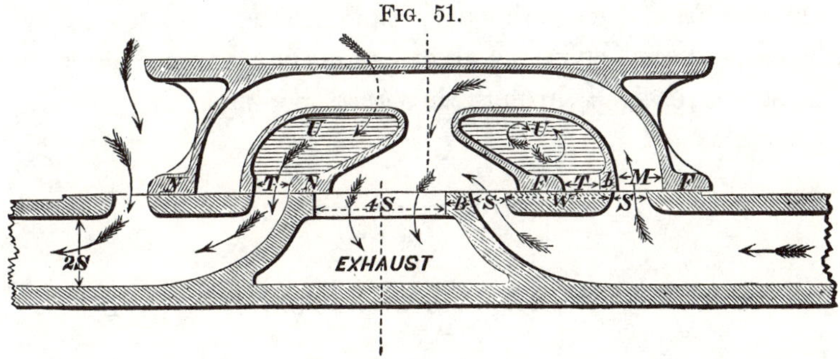

Fig. 51.

attained by increasing the number of the steam ports two or three fold, so that $\frac{1}{2}$ or $\frac{1}{3}$ of the travel will suffice to open the same extent of port area. For a reduction in the travel of one-half, the parts are commonly arranged as shown in Fig. 51.   In this the valve edges F and N are separated by a distance wide enough to admit two steam passages U, U, whose openings T T communicate with the inner ports but not with the exhaust.   They are formed with a width equal to that designed for the port opening.   The general proportions of such valves may be concisely expressed as follows:

Total width of steam passage=2 S.
Each port=S.   Width of port opening=T.
Exhaust port=4 S.
Valve faces F and N=S+lap of valve.
Bridge W=$\frac{1}{2}$ travel+lap+T+$\frac{3}{4}$ inch.
Width of exhaust bridge B dependent on thickness of cylinder.
Width of exhaust bridge b dependent on thickness of valve.

Occasionally the travel is reduced still further by inserting a third port beyond the other two.   When so arranged the outer faces F and N are extended and through them passages cut, which admit the steam to the cylinder, but remain closed during the period of exhaust; consequently the *two* exhaust passages must be amply large to discharge the steam received through the *three* openings.

LINK AND RECIPROCATING MOTION COMBINED.

It was observed, while discussing the subject of link motion, that the central line of motion might be inclined at any angle to that of the piston motion, without affecting the character of the valve action.   Fig. 52 illustrates this pecu-

liarity (for one position, namely, an angle of 75°) as applied
to a back-action engine having an independent cut-off.
The line E F represents the central line of the link motion,

FIG. 52.

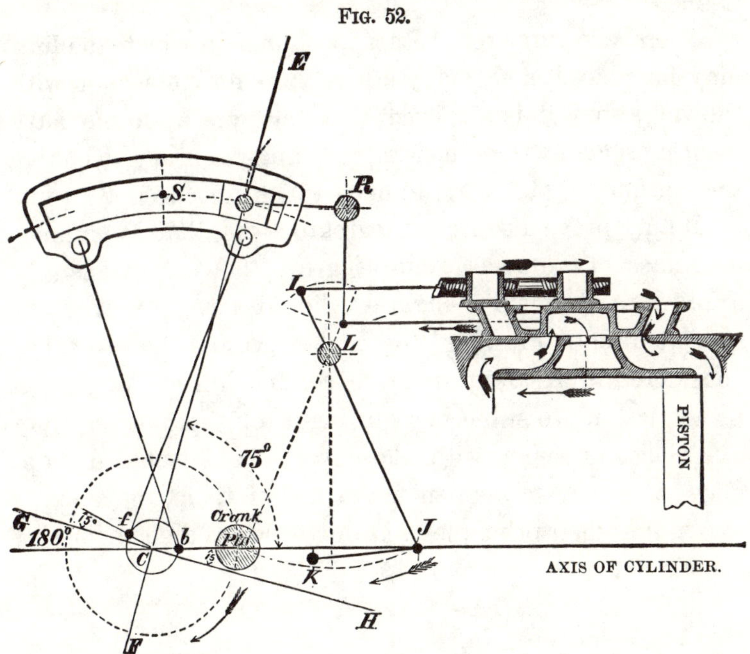

and G H a line at right angles thereto from which are laid
off the 15 degree angular advances of the eccentrics. The
eccentric rods being crossed, the lead of course diminishes
from the full to the mid gear of the link, but as this is only
employed in the full or immediately adjacent gears, the reduction proves advantageous, since it enables the engineer
to stop his engine by placing the link in the mid gear (see
Fig. 26). The cut-off valves are operated by a lever I J
connected through the link J K with the pin K, which is
secured to one of the piston rods or to the cross-head in a
direct-action engine.

This valve motion might have been derived from an eccentric with its normal position in the line E F, and acting on the valve through the medium of a rocker.   The general conditions governing the motion have already been examined.

There are numerous other positions in which the link may be placed, and many other plans for connection with the valve, but it is believed that the typical forms have been presented with sufficient care and accuracy to enable the designer by their aid to accomplish any desired result.

It may prove a source of regret to some, that the numerous class of automatic cut-off gears, now so extensively applied to stationary engines, should have received no special attention in this Work.   The Author however, has considered it most expedient to confine himself to general principles and to subjects requiring a solution in every day experience, leaving with those who hold such monopolies, and who *alone* can make use of them, the onus of explaining their principles and advertising their points of excellence.

APPENDIX.

FORMULÆ RELATING TO CRANK AND PISTON MOTIONS.

CRANK ANGLE EQUATIONS.

Upon comparing the positions assumed by a piston under the influence of a connecting rod of virtually infinite length, with those resulting from the use of a rod of finite length; it appears that for any point selected in the forward stroke (as e. Fig. 53)

FIG. 53.

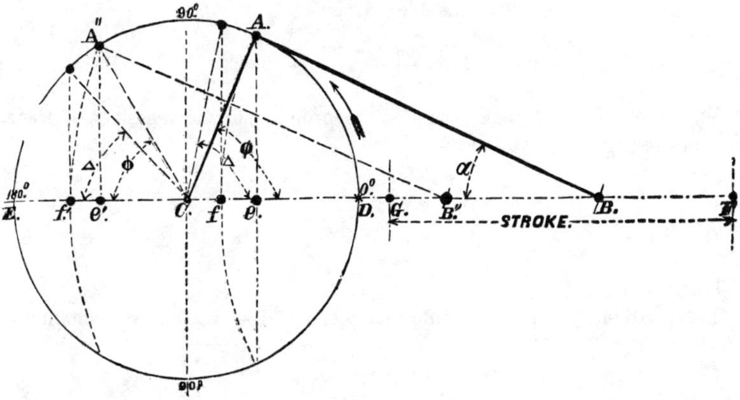

the piston is advanced beyond its normal position by a distance ef equal to the ver. sine of the angle a included between the connecting rod and centre line of the engine. On the return stroke it will be found in arrear by the same distance.

The object of this investigation is to determine formulæ for ascertaining the crank angle due to a certain length of connecting rod and position of the piston.

Let the length of the crank arm A C (in inches)$=k$ and that of the connecting rod A B (in inches) $= G$; their ratio will be that of $k : G$; but if $\dfrac{G}{k} = N$, the ratio of the crank arm to the connecting rod will become $1 : N$.

Again the distance of the piston from the commencement of the stroke (in inches) $= f$, consequently :—

$$\cos \Delta \begin{cases} \text{for an advance } less \quad \text{than } \tfrac{1}{2} \text{ stroke} &= \dfrac{k-f}{k} \\[2ex] \text{``} \qquad \text{``} \qquad greater \text{ `` } \tfrac{1}{2} \text{ ``} &= \dfrac{f-k}{k} \end{cases}$$

In which the angle Δ is a location the crank would assume, if the ratio of crank arm to connecting rod were that of $1 : \infty$.

Considering ϕ as the correct angle for the crank, when the ratio is that of $1 : N$ and its advance less than the first half stroke, we have :—

$$Ce = k . \cos \phi.$$

$$fB = G.$$

$$eB = G . \cos a$$

but,
$$Ce = fB - eB + Cf.$$

or
$$k \cos \phi = G - G \cos a + k \cos \Delta.$$

Dividing both members by k

$$\cos \phi = N - N \cos a + \cos \Delta \tag{1.}$$

or,
$$\cos \phi - \cos \Delta = N (1 - \cos a).$$

$$\therefore \quad \cos a = 1 - \frac{\cos \phi - \cos \Delta}{N} \tag{2.}$$

Since the " sides of plane triangles are proportional to the sines of their opposite angles " it follows :—

$$N : 1 :: \sin \phi : \sin a$$

$$\therefore \quad \sin a = \frac{\sin \phi}{N}$$

Transforming
this equation upon the principle that $\cos^2 a = 1 - \sin^2 a$. We obtain :—

$$\cos a = \sqrt{\left(1 - \frac{\sin^2 \phi}{N^2}\right)}$$

Substituting this value of $\cos a$ in Eq. No. 2 gives

$$\sqrt{\left(1 - \frac{\sin^2 \phi}{N^2}\right)} = 1 - \frac{\cos \phi - \cos \Delta}{N}$$

Squaring both members of the equation ;—

$$1 - \frac{\sin^2 \phi}{N^2} = 1 - 2 . \frac{\cos \phi - \cos \Delta}{N} + \frac{\cos^2 \phi - 2 \cos \phi \cos \Delta + \cos^2 \Delta}{N^2}$$

and multiplying by N^2 :—

$$N^2 - \sin^2 \phi = N^2 - 2 N \cos \phi + 2 N \cos \Delta + \cos^2 \phi - 2 \cos \phi . \cos \Delta + \cos^2 \Delta$$

    cancelling and changing signs.

$$\text{Sin}\,^2\phi = 2\,\text{N}\cos\phi - 2\,\text{N}\cos\Delta - \cos\,^2\phi + 2\cos\phi.\cos\Delta - \cos\,^2\Delta$$

or

$$(\sin\,^2\phi + \cos\,^2\phi) - 2\,\text{N}\cos\phi - 2\cos\phi.\cos\Delta = -\,2\,\text{N}\cos\Delta - \cos\,^2\Delta$$

which reduces to :—

$$-\,1 + 2\,\text{N}\cos\phi + 2\cos\phi.\cos\Delta = 2\,\text{N}.\cos\Delta + \cos\,^2\Delta$$

or $\quad 2\,(\text{N} + \cos\Delta)\cos\phi = 2\,\text{N}\cos\Delta + (1 + \cos\,^2\Delta)$

$$\therefore\cos\phi = \frac{\text{N}\cos\Delta + (\tfrac{1}{2}\cos\,^2\Delta + 0.5)}{\text{N} + \cos\Delta} \tag{3.}$$

When the piston stands at the half stroke point, the angle $\Delta = 90°$. But since the $\cos 90° = 0$. the equation (3.) will be so modified as to read :—

$$\cos\phi = \frac{0.5}{\text{N}} \tag{4.}$$

For an advance greater than the half stroke the equation No. 1. will read thus :—

$$\cos\phi = \cos\Delta - \text{N} + \text{N}.\cos.\,a$$

which followed through similar transformations to those that developed equation No. 3. gives finally :

$$\cos\phi = \frac{\text{N}\cos\Delta - (\tfrac{1}{2}\cos\,^2\Delta + 0.5)}{\text{N}-\cos\Delta} \tag{5.}$$

If instead of the total crank angle or its supplement, the irregularity due to the connecting rod is required, the question is reduced to finding the difference between $\cos\Delta$ and $\cos\phi$.

Subtract $\cos\Delta$ from both members of equation No. 3.

$$\cos\phi - \cos\Delta = \frac{\text{N}\cos\Delta + (\tfrac{1}{2}\cos\,^2\Delta + 0.5)}{\text{N} + \cos\Delta} - \cos\Delta$$

Let E = the irregularity, then :—

$$\text{E} = \frac{\text{N}\cos\Delta + \tfrac{1}{2}\cos\,^2\Delta + 0.5 - \text{N}\cos\Delta - \cos\,^2\Delta}{\text{N} + \cos\Delta}$$

Multiplying by $\tfrac{2}{2}$ and cancelling :—

$$\text{E} = \frac{1 - \cos\,^2\Delta}{2\,(\text{N} + \cos\Delta)}$$

$$\therefore\text{E} = \frac{\sin\,^2\Delta}{2\,(\text{N} + \cos\Delta)} \tag{6.}$$

and similarly for an advance greater than $90°$.

$$\text{E} = \frac{\sin\,^2\Delta}{2\,(\text{N} - \cos\Delta)} \tag{7.}$$

With equations 6 and 7 also the term N being given, the irregularity E can be determined for all values of Δ. The latter may be combined, to form a scale, by erecting them as ordinates to a straight line and joining their extremites. The curved line so described will be a species of the cycloid.

The equations 3 and 5 can be modified by making the original values of cos Δ,—

$$\frac{k - f}{k} = C. \qquad \text{and} \qquad \frac{f - k}{k} = e.$$

read thus :—

For cut-off *less* than $\frac{1}{2}$ stroke. $\cos \phi = \dfrac{N.C + (\frac{1}{2}c^2 + 0.5)}{N + C}$ (8)

"   "   *equal* to $\frac{1}{2}$ " $\cos \phi = \dfrac{0.5}{N}$ (9)

"   "   *greater* than $\frac{1}{2}$ " $\cos \phi = \dfrac{N.e - (\frac{1}{2}e^2 + 0.5)}{N - e}$ (10)

General formulæ for finding that crank angle, which corresponds with a given position of the piston.

It will be observed that equation No. 8 applies to all piston advances less than the half stroke in the forward motion, and greater than the same in the return motion; while No. 10, on the contrary, solves those greater than the half stroke in the forward and less in the return motion.

EXAMPLES.

No. 1.   The stroke of a certain engine = 32″.
         The cut-off takes place at   20″.
         Ratio of crank arm to connecting rod 1 : 5.

Required.
         The angles of the crank at the moments of cut-off.

Since the cut-off is greater than the half stroke

$$e = \frac{f - k}{k} = \frac{20 - 16}{16} = 0.25$$

For the forward stroke equation No. 10

$$\cos \phi = \frac{Ne - (\frac{1}{2}e^2 + 0.5)}{N - e} = \frac{1.25 - (.03125 + 0.5)}{5 - 0.25} = 0.1513$$

And $\therefore$ $\phi = 81°. 18'$.

       Supplement of $\phi = 98°. 42' = $ crank angle forward.

Again, substituting N and *e* in equation No. 8 gives :—

$$\cos \phi = \frac{5. \times 0.25 + (0.03125 + 0.5)}{5 + 0.25} = 0.3393$$

searching in a table of natural cosines for this decimal, we find that :—

$$\phi = 70°. 10' \quad \text{and}$$

supplement of $\phi = 109°. 50' =$ crank angle of the return stroke, or the second of the two angles sought.

They have both been given in the STROKE TABLE column for the ratio 1 : 5, and line for 0.625 piston position.

No. 2. Stroke of engine $= 10' \, 6''. = 126.''$

Cut-off of steam at $33''\frac{1}{2}$.

Ratio of crank arm to connecting rod $= 1 : 7\frac{4}{10}$.

Required.

The angles of the crank at the moment of cut off.

The cut-off being less than one-half the stroke

$$c = \frac{k - f}{k} = \frac{63 - 33.5}{63} = 0.46826$$

This substituted in equation No. 8 gives :—

$$\cos \phi = \frac{7.4 \times .46826 + (.10965 + .5)}{7.4 + .468} = 0.518$$

And from a table of natural cosines :—

$$\phi = 58°. 49' = \text{crank angle forward.}$$

Proceeding in like manner for the return stroke with equation No. **10**

$$\cos \phi = \frac{3.46512 - (.10965 + .5)}{7.4 - .468} = 0.4119.$$

$$\therefore \quad \phi = 65°. 40' = \text{crank angle of the return stroke.}$$

No. 3. What irregularity is impressed on a valve at its half travel by a ratio between the eccentric throw and length of eccentric rod of 1 : 25. ?

$$\text{From Equation No. 4, } \cos \phi = \frac{0.5}{N} = \frac{0.5}{25} = 0.02$$

$$\therefore \quad \phi = 88°. 51'. \quad \text{and the}$$

Irregularity (or complement of ϕ) $= 1°. 9'.$ about $1°\frac{1}{8}$, which for **any given** travel may be measured from the Base Line of the TRAVEL SCALE.

The irregularity for many other ratios is given below :–

Irregularity for ratio		Ratio	
1 : 40 = $\frac{3}{4}°$		1 : 8 = $3°\frac{5}{8}$	
" " 1 : 30 = $\frac{7}{8}°$		1 : 7 = $4°\frac{1}{8}$	
" " 1 : 20 = $1°\frac{3}{8}$		1 : 6 = $4°\frac{3}{4}$	
" " 1 : 15 = $1\frac{7}{8}°$		1 : 5 = $5°\frac{3}{4}$	
" " 1 : 12 = $2°\frac{3}{8}$		1 : $4\frac{1}{2}$ = $6°\frac{3}{8}$	
" " 1 : 10 = $2\frac{7}{8}°$		1 : 4 = $7\frac{1}{4}°$	

No. 4.   When a valve, having an eccentric throw and rod ratio of 1 : 6 attains its quarter stroke (30°) point, by how many degrees does it stand in advance of the position it would occupy were the throw and rod ratio that of 1 : ∞ ?

Referring to equation No. 7. we observe that in this case $\Delta = 60°$. and as; $\cos 60° = \frac{1}{2}$ $\sin. 60° = \frac{1}{2}\sqrt{3}$ and $\sin{}^2 60° = \frac{3}{4}$ the equation for E will become :

$$E = \frac{\frac{3}{4}}{2N - 1} = \frac{3}{8N - 4.}$$

Also, $\cos \phi = \cos \Delta - E$, or $\phi = \cos{}^{-1} (.5 - E)$

Since the angular correction $= \phi° - \Delta°$.
we have   "   " $= \cos{}^{-1} (5 - E) - 60°$.

By substituting 6 the value of N in equation for E, we obtain :—
Correction $= \cos{}^{-1} (5 - .0682) - 60°$.
" $= 64°. 25'. - 60°$
" $= 4°. 25' =$ amount in advance.

In a similar manner, we can determine the extent of the irregularity, which may be expected from other ratios, at the quarter stroke or 30° angular advance point, and measure their linear value from the 30° line on the Travel Scale.

Ratio 1 : 30	between eccentric throw and rod, gives;				$\frac{3}{4}°$
" 1 : 25	"	"	"	"	$1°$
" 1 : 20	"	"	"	"	$1\frac{1}{4}°$
" 1 : 15	"	"	"	"	$1\frac{3}{4}°$
" 1 : 10	"	"	"	"	$2\frac{5}{8}°$
" 1 : 8	"	"	"	"	$3\frac{1}{4}°$
" 1 : 7	"	"	"	"	$3\frac{3}{4}°$
" 1 : 6	"	"	"	"	$4\frac{3}{8}°$
" 1 : 5	"	"	"	"	$5\frac{3}{8}°$
" 1 : 4	"	"	"	"	$6\frac{7}{8}°$

INDEX.

30. T.

5 9 3 5 6 4/8